Marketing Intelligent Design

Recently, a new battle has emerged between science and religion. The battle focuses on intelligent design (ID) and the numerous legal, philosophical, and educational concerns surrounding it. Resolution of these concerns centers on two questions: Is ID science? Is ID religion? Despite the fact that ID does not meet the standards of scientific rigor, ID proponents have been able to create a remarkably well-designed marketing plan aimed at imposing a theistic naturalism in schools and scientific discourse. Both the ID movement and some of its most vociferous opponents have a vested interest in suggesting that science, especially evolutionary biology, and religion are incompatible. This book presents a legal and philosophical counterpoint by demonstrating the compatibility between religion and evolutionary biology and the incompatibility between ID and mainstream science.

Professor Frank S. Ravitch is Walter H. Stowers Chair in Law and Religion at the Michigan State University College of Law. He is the author of, most recently, *Masters of Illusion: The Supreme Court and the Religion Clauses* (2007) and *Law and Religion, A Reader: Cases, Concepts, and Theory*, 2nd edition (2008). He also has published numerous law review articles dealing with law and religion, civil rights, and disability discrimination. Ravitch regularly serves as an expert resource for print and broadcast media and speaks on topics related to church–state and civil rights law to a wide range of national and local organizations.

Marketing Intelligent Design

Law and the Creationist Agenda

FRANK S. RAVITCH

Michigan State University College of Law

CAMBRIDGE UNIVERSITY PRESS
Cambridge, New York, Melbourne, Madrid, Cape Town, Singapore,
São Paulo, Delhi, Dubai, Tokyo, Mexico City

Cambridge University Press
32 Avenue of the Americas, New York, NY 10013-2473, USA

www.cambridge.org
Information on this title: www.cambridge.org/9780521139267

First published 2011

Printed in the United States of America

A catalog record for this publication is available from the British Library.

Library of Congress Cataloging in Publication data

Ravitch, Frank S., 1966–
Marketing intelligent design : law and the creationist agenda /
Frank S. Ravitch.
p. cm.
Includes bibliographical references and index.
ISBN 978-0-521-19153-1 (hardback)
1. Creationism – Study and teaching – Law and legislation – United States. 2. Evolution (Biology) – Study and teaching – Law and legislation – United States. 3. Intelligent design – Study and teaching – United States. 4. Creationism – Study and teaching – United States. 5. Evolution (Biology) – Study and teaching – United States. 6. Religion in the public schools – Law and legislation – United States. I. Title.
KF4208.5.S34R38 2010
231.7′652 – dc22 2010015167

ISBN 978-0-521-19153-1 Hardback
ISBN 978-0-521-13926-7 Paperback

This book is dedicated to my wife, Jamie,

my daughters, Elysha and Ariana,

and my parents, Carl and Arline.

Thank you for your love and support.

Contents

Preface

This book and the articles I have published on intelligent design (ID) did not begin as a research project aimed at demonstrating the scientific, philosophical, and legal problems with treating ID as science. For years, I, along with many other people, have been fascinated by the debate over evolution. From my vantage point, I never saw any inherent inconsistency between scientific evidence that overwhelmingly supports evolution and my faith. With the caveats set forth later in this book, I still do not. Yet I initially wondered if ID would be a plausible alternative to traditional evolutionary biology.

After reading much of what ID advocates have written, I realized that ID is not a plausible alternative to evolution, at least not a scientifically plausible alternative. In fact, the more ID material I read (i.e., material written by ID advocates), the more convinced I became that ID is a marketing strategy based on religious apologetics and not truly a scientific approach. This realization had the potential to disappoint, but, in the process, I gained a better understanding of my own religious commitments and the relationship between these commitments and science. For me, however, ID has no role in either. After all, it is hard to base one's faith or one's understanding of the natural world on a marketing strategy, albeit an exceptionally well-designed marketing strategy.

Having spent many years studying law and religion, I am familiar with the religious apologist tradition of trying to prove the truth of faith commitments. More specifically, I am familiar with the idea of natural theology and, specifically, of the form that that idea took in the writings of the Reverend William Paley and his contemporaries. Reading the work of ID advocates, it was immediately obvious to me that much of their work was just a rehash of Paley's ideas, disguised in fancy terminology or brought into the realm of biochemistry.

Moreover, because I have spent much time working with interpretive theory, especially philosophical hermeneutics, I am familiar with a variety of philosophical perspectives that can be applied to the sciences. Specifically, I am familiar with the concept of relativism. It struck me that ID advocates have no choice but to rely on relativism if they wish to have their work accepted under any rubric of science. Of course, the reason for this is that ID advocates would have to redefine science for ID ever to be accepted as science. To do so in an intellectually consistent fashion, however, ID advocates must argue that science is relative. This means that almost anything could count as science. In fact, I saw these types of arguments in some of the legal and philosophical arguments made by ID advocates, but the authors never seemed to grapple with the implications of using relativism.

Another interesting facet to ID is that ID advocates do not acknowledge that the designer is God. In considering this aspect as a law and religion scholar, it was difficult to miss the legal implications. I began to wonder about the relationship between the legal principle that religion cannot be taught as science in a public school science classroom and the evolution of ID (especially the ID movement's general denial that ID is religious and that the designer is God). This question alone would have been worth the time to engage in this subject, but the project has taught me so much more and has piqued my interest in the relationship between law, religion, and science.

Acknowledgments

I owe a great debt to many of my colleagues both at Michigan State University (MSU) and around the world for their support and comments. Most important, I am grateful to John Berger of Cambridge University Press for supporting this project, for his great sense of humor, and for his tireless efforts to shepherd this book to press in a careful yet timely fashion. I also would like to thank Lane Dilg, Stephen Feldman, Chip Lupu, Ken Marcus, William Marshall, Bob Tuttle, and William Van Alstyne for comments received at a symposium sponsored by the WILLIAM AND MARY BILL OF RIGHTS JOURNAL on an article that became part of Chapter 6. Many thanks also go to Kristi Bowman and Mark Modak-Truran, with whom I served on a panel at the Law and Society Association 2008 Annual Meeting, where I presented some of the material that became part of Chapters 2 and 3.

I am also grateful to Steve Goldberg, Kevin Saunders, Brett Scharffs, Cynthia Starnes, and Glen Staszewski for their comments and support.

I would like to thank my dean, Joan Howarth, for her consistent support for this project. Thanks also to Charles Ten Brink, Jane Edwards, Hildur Hanna, and the staff of the MSU law library for their consistently excellent support; to Amanda Gardiner, Kara Jansen, Holly Shannon, and Andrew Walter for excellent research assistance; and to my secretary, Jacklyn Beard. Moreover, I would like to thank the anonymous readers who reviewed the book proposal and the lengthy articles attached to it for their helpful comments. Of course, any errors are mine alone.

The term *acknowledgments* is inadequate to express my gratitude to my wonderful family for their constant love and support: first of all, to my wife, Jamie, who is supportive of my work even when she disagrees with me; and to my amazing and wonderful daughters, Elysha and Ariana, whose energy, intelligence, and love never cease to amaze me and who make life brighter and make me smile even on the busiest of days. My parents, Arline and Carl Ravitch, are the best parents any child could hope for; without them, none of the things I have been fortunate enough to do would ever have been possible. They are a constant source of love, support, joy, and inspiration. I cannot forget my late Bubby and Pop-Pop, who, even after their passing, continue to inspire me and serve as examples of hard work, love, and kindness. I also thank Barbara and Gerry Grosslicht, who defy any negative stereotype of in-laws and who are always caring and supportive. My sisters, Elizabeth and Sharon, and my sister-in-law, Karen, and their wonderful families are not only supportive but also wonderful people. Thank you to my grandmother-in-law, Hilda Sorhagen, who is fascinated by questions of religion and is always interested in what I am doing. Last, but not least, I thank my Uncle Gary and Aunt Mindy and Aunt Jackie and Uncle Ken and their wonderful families, who also are interested in whatever I do.

As mentioned earlier in these acknowledgments, I have incorporated modified sections of two articles into this book: *Playing the Proof Game: The Law and Intelligent Design*, 113 PENN ST. L. REV. 841 (2009) and *Intelligent Design in Public Universities: Establishment of Religion or Academic Freedom*? 16 WILLIAM & MARY BILL RTS. J. 1061 (2008) (symposium).

1 Designing Design

> Design theory promises to reverse the stifling dominance of the materialist worldview, and to replace it with a science consonant with Christian and theistic convictions.[1]

> Concepts concerning God or a supreme being of some sort are manifestly religious. . . . These concepts do not shed that religiosity merely because they are presented as a philosophy or as a science.[2]

I. Introduction to Intelligent Design

Every day in public schools, universities, houses of worship, and coffee shops, a battle rages over where humanity came from or, more specifically, how humans came to be human. Much of the debate is focused on whether a supposedly new concept of human origins – intelligent design (ID) – should be taught in public schools. Yet few people know much, if anything, about this concept: how it came to the fore, and what it means for law, science, faith, and the future of America.

ID advocates have a vested interest in this confusion. ID is, at least partially, a response to the success of modern science, especially evolutionary biology and cosmology, in explaining natural phenomena. Yet the form ID has taken is primarily an attempt to respond to several important cases decided by the U.S. Supreme Court[3] and to win in the court of public opinion – not exactly an auspicious baseline for a so-called scientific theory. This book explains that the essence of ID lies in a solid marketing plan and an attempt to avoid legal constraints, not in promoting a serious scientific alternative to evolutionary biology and biochemistry.

Since the 1960s, attempts to teach creationism and "creation science," as well as attempts to exclude or marginalize evolution in public school science classrooms have been found unconstitutional.[4] At the same time, science has made remarkable strides in explaining numerous natural phenomena and in fostering a technological explosion. These developments have led to the increased prevalence of support for what ID advocates often refer to as *scientific materialism* – basically a fancy term for the notion that natural phenomena and human behavior are the products of natural forces that can be explained by science.[5] As seen in later chapters, scientific materialism need not conflict with religion or faith, and it is the ID movement itself that tries to promote the idea that there is an inherent conflict between scientific materialism and faith.

ID is more about marketing creation in a manner that will enable it to be taught in the public schools and accepted in public discourse than it is about scientific disagreement.[6] This is why ID advocates rarely acknowledge that the intelligent designer is God. It is also why confusion is one of ID's greatest weapons.

Many people of faith believe that God must have had some role in the complexity we see in the universe. This is, however, inherently theistic and therefore problematic when introduced as science in public schools. Still, numerous people of faith believe in what can loosely be called *theistic evolution* – quite simply, the notion that the scientific proof for evolution is so overwhelming that it would be ludicrous to ignore it, but that this in no way precludes a belief that God created life.[7] Evolution may simply be the mechanism that God used.[8]

For now, it is enough to note that from the perspective of theistic evolution, there is no reason to teach theistic views of human origins in science classrooms or to attempt to view theistic concerns through the lens of science. This is because theistic evolutionists accept modern science and do not see it as inconsistent with faith – faith is faith, not science. Theistic evolution is explored in greater depth in Chapter 4. Conversely, intelligent designers seek to explain the existence of the designer through what they argue is science, an argument that is at the core of the issue.[9] This has significant legal ramifications because it causes ID proponents to enter into what I have called the *proof game*.[10]

If ID advocates had simply proposed their ideas in a philosophical or theological context – ideas that are already thousands of years old in those disciplines[11] – there would be little dispute. After all, in a free society, there is nothing wrong with believing in design. The problem arises when ID enters the proof game in the scientific context. The movement has a vested interest in doing this so that it can market its ideas in science classrooms and appeal to public sentiment,[12] but to be taken seriously in scientific circles or taught in public school classrooms legitimately and without violating the Constitution, ID must be science.[13] Thus, the proof game is everything to ID proponents.

By couching ID as science and not theology, ID proponents are able to argue for access to the forum of scientific debate. As will be seen, they often treat the scientific realm as a forum for debate of philosophical and metaphysical questions. Such questions, however, are not generally answered by studying the natural world – the world to which modern science is devoted.[14] ID advocates then claim that ID is being discriminated against when it is excluded from the scientific forum.[15] These claims seem to be based in free speech concepts, often cast by ID proponents in broad terms such as "academic freedom" and "fairness."[16] Such arguments are, however, question begging.

If ID is a scientific theory, it could have a place in scientific discourse, but if not, the ID movement's rhetoric of exclusion and discrimination is nothing more than crying wolf. Otherwise, alchemy could claim a place in chemistry classrooms because excluding it would be discrimination. The same would be true for astrology in astronomy classes and UFOlogy in a number of fields.[17] As will be seen, from a legal perspective the ID movement's arguments about exclusion suggest that public school classrooms and scientific research are limited public fora, that is, fora opened to debate about all ideas claimed to be scientific, no matter how discredited.[18] As explained in Chapter 3, this argument makes no sense legally or philosophically.

Under current legal standards, the ID movement must redefine science to justify including ID in science courses.[19] In the *Kitzmiller* case, discussed in depth in Chapter 3, Michael Behe, a biochemist who is also a leading proponent of ID theory, acknowledged under intense questioning that a definition of science that includes ID would also include astrology.[20] In all fairness to Behe, he had no choice because there is no way around this conundrum when one tries to include ID within the definition of science. The key for present purposes is that the definition of science is so important to ID proponents precisely because of the law surrounding the teaching of human origins in public schools and universities.

In response to these concerns, ID proponents often raise the specter of *secular humanism* and *scientific materialism*.[21] They argue that evolutionary biology privileges secular humanism and a materialistic worldview and excludes all alternatives.[22] They claim that ID provides a counterbalance to the establishment of secular humanism in the public schools.[23] Yet they do so without arguing that ID is a religious alternative; rather, they argue that ID is an alternative to *scientific materialism* and *methodological naturalism*, as they define those concepts.[24] When the history and tenets of ID are considered, this amounts to the same thing as openly acknowledging ID as a religious concept. Moreover, what ID proponents call secular humanism is really just plain secularism.[25]

Secular humanism is a philosophy and actually shares some common ground with religious humanism.[26] The organized followers of secular humanism often follow the teachings of the *Humanist Manifesto*,[27] which is a document that sets forth the tenets of secular humanism, at least for those who are part of the secular humanist groups that adhere to it. This should not be confused with secularism more generally, which comprises essentially the nonreligious. Science is secular. Science studies the natural world and sometimes leads to discoveries that benefit people in important ways, but it does not have a humanist agenda or moral philosophy.[28] Thus the use of the term *secular humanism* by ID advocates evokes a straw man in an effort to affect public consciousness. This has serious implications.

None of this means that the ID movement is wrong to assert that much commentary on evolutionary biology, especially by scientists, promotes naturalism and materialism at the expense of supernatural explanations. Many scientists have ventured far beyond their scientific areas of expertise to express opinions on what evolutionary theory suggests about God. Perhaps most notable among this cadre is Richard Dawkins,[29] whose work is discussed in depth in Chapter 5, but he is joined by many others, including geneticist Richard Lewontin and scientific philosopher Daniel Dennett.[30]

Of course, this does not mean that by its nature, evolutionary biology leads to such conclusions; rather, as will be explained in Chapter 5, it simply means that it has been applied by some in such a manner.[31] As discussed in Chapter 4, many religious leaders, theologians, and religious movements,[32] along with some leading scientists,[33] have explained that evolution and religion need not conflict with each other. ID advocates assume that naturalism and materialism inherently conflict with religion,[34] as do some of their most vociferous opponents,[35] but as biochemist Kenneth Miller has explained, this is not inevitable or even logical.[36]

Evolution is part of science and as such, it is testable and helps explain the natural world.[37] It has been immensely successful in this regard.[38] Yet one must go beyond the scientific realm to address metaphysical questions. To the extent ID advocates and some of their opponents venture into this realm, they venture beyond science (although, as seen in Chapter 5, Richard Dawkins makes some powerful arguments concerning the role probabilities may play in assessing aspects of the metaphysical realm).[39] The answer to ID advocates' claim that modern science, especially evolution, is a threat to religion cannot be found by asserting that ID is science, unless, of course, that can be proven, but rather by acknowledging that some of the brilliant scientists and philosophers who have pitted evolution against religion are not good theologians.

Another facet of the ID debate involves a persecution complex that many ID advocates seem to have internalized and in which legal conceptions play a significant role. In a recent movie called *Expelled: No Intelligence Allowed* (2008),[40] Ben Stein suggests that ID advocates are being persecuted in the educational and scientific arenas and that this persecution conflicts with free speech and intellectual fairness.[41] Similar arguments have been made by a number of ID proponents.[42] Yet there are standards and law that relate to what can and cannot be done in academic contexts, and as with most things, the story of these expulsions told by Stein and others leaves out many salient and important facts.[43] Surely Stein raises some important questions about academic and scientific discourse, but as seen in Chapter 6, the answers are not necessarily what Stein and other ID proponents imply.[44]

II. A Basic Primer on Creationism, Creation Science, and the Supreme Court

On July 20, 1981, Louisiana Governor David C. Treen signed the Balanced Treatment for Creation-Science and

Evolution-Science in Public School Instruction Act into law. The law was sponsored by state senator Bill Keith, who introduced a related bill in June 1980.[45] The stated purpose of the law was to promote academic freedom,[46] but it did so by requiring that "creation science" be taught whenever evolution is taught in Louisiana public schools.[47] There was no explicit prohibition on teaching creation science before the law was enacted,[48] and under the law, there was no requirement that either creation science or evolution be taught.[49] The only requirement was that teachers must teach creation science if they teach evolution.

The Louisiana law was an example of what came to be known as balanced treatment laws. These laws were supported by the creation science movement, which existed long before the current ID movement.[50] Creation science evolved mostly from a sometimes uncomfortable pairing of *old Earth creationists* and *young Earth creationists*.[51] Old Earth creationists believe the Earth may be quite old but that complex life-forms – especially human beings – were placed here by God in their present form.[52] Young Earth creationists take the time line in the Bible literally and date the creation of the Earth and humanity to about six thousand years ago.[53] A few young Earth creationists would allow for a slightly older Earth, but even these people would suggest the age to be in the thousands of years, not millions, and certainly not billions, of years.[54]

Interestingly, the creation science movement, like the ID movement, was designed to gain public acceptance of creationism and especially to gain access to science classes in the public schools.[55] By couching creationism in scientific terms, creation scientists hoped to win the legal battle over the constitutionality of teaching creation science in the public schools. One of the major strategies creation science advocates employed was balanced treatment laws like the one in Louisiana.[56] Creation scientists argued that these laws were designed to promote academic freedom and free speech.[57] The Louisiana law was challenged in federal court shortly after it was signed.[58] The resulting decision issued by the Supreme Court in 1987 is known as *Edwards v. Aguillard*.[59]

In *Edwards*, the Court held that the Louisiana law was unconstitutional because its purpose was to promote a religious concept, creation science, and not to promote academic freedom.[60] *Edwards* was a major defeat for the creation science movement and a defining moment for what would become the ID movement.

The *Edwards* Court focused exclusively on whether the Louisiana Balanced Treatment Act had a valid secular purpose.[61] After looking at the language of the Louisiana Balanced Treatment law[62]; the statements of Senator Keith, who introduced it;[63] statements by other legislators and government officials;[64] and statements by those who testified before the legislature on the bill,[65] the Court held that the purpose of the law was to promote creationism and to favor the views of certain Christian denominations.[66] The Court did not accept the state's argument that the law was designed to promote academic freedom.[67] It found instead that the law could not serve the purpose of promoting academic freedom because the law limited rather than expanded such freedom.[68]

Weighing heavily against the claim of a valid secular purpose were that the law's proponents spoke in explicitly religious terms and that creation science posits that human beings were placed on Earth by a supernatural creator.[69] Moreover, the Court found that the law was designed to counter evolution with creationism "at every turn,"[70] which served the beliefs of certain religious groups.[71] Further undermining the state's argument was that the law provided support for additional creation science teaching materials but not for the development of additional evolution materials,[72] and the law explicitly provided protection for teachers who taught creation science but not for those who taught evolution (even though if one was taught, the other had to be taught under the law).[73]

Perhaps most important, the Court rejected the argument that because creation science marketed itself as science, it did not promote religion.[74] The following quote from *Edwards* is particularly relevant to the ID debate:

> The preeminent purpose of the Louisiana Legislature was clearly to advance the religious viewpoint that a *supernatural being* created humankind. . . . Senator Keith's leading expert on creation science, Edward Boudreaux, testified at the legislative hearings that the theory of creation science

included belief in the existence of a *supernatural creator*.[75] (italics added)

Significantly, the *Edwards* decision, and the defeat of balanced treatment acts in other cases,[76] became an impetus for what would become the ID movement.[77] In fact, when one looks at the basic tenets of the ID movement, it seems clear that ID was designed in part to avoid some of the problems that doomed creation science in the courtroom.[78] After all, a major goal of the ID movement is to introduce ID in the public schools[79] or, at the very least, to introduce curricular changes that call evolution into question.[80] The grander plans of the ID movement will be less likely to succeed if ID cannot gain access to public school classrooms and win the hearts and minds of future scientists and philosophers of science.

This is a bit ironic because creation science was itself a response to earlier legal defeats suffered by laws that promoted creationism by requiring that creationism be taught, prohibiting the teaching of evolution, or both.[81] The most notable of these earlier cases, *Epperson v. Arkansas*,[82] was decided by the U.S. Supreme Court in 1968. In that case, the Court held that an Arkansas law prohibiting the teaching of evolution in its public schools and universities violated the Establishment Clause.[83]

The Arkansas law made it a crime to teach evolution in public schools and universities and exposed any teacher who did so to being fired.[84] The Little Rock School District recommended a new biology text in 1965, which included instruction on evolution.[85] A young biology teacher in the district, Susan Epperson, realized that if she taught the evolution section in the new book, she would potentially be subject to dismissal and criminal liability under the state law, even though the school district had approved the text.[86] She believed that teaching the material on evolution was in the best interest of her students, and she sued, asking the courts to declare the law unconstitutional and therefore unenforceable against her and others.[87]

The Court agreed with Epperson.[88] It held that the law was unconstitutional because it did not have a secular purpose.[89] The Court found that the Arkansas law was designed to prevent evolution – and only evolution – from being taught in the public schools because evolution was antithetical to a particular religion:

> [T]here can be no doubt that Arkansas has sought to prevent its teachers from discussing the theory of evolution because it is contrary to the belief of some that the Book of Genesis must be the exclusive source of doctrine as to the origin of man. No suggestion has been made that Arkansas' law may be justified by considerations of state policy other than the religious views of some of its citizens.[90]

Moreover, the law could not be defended on the ground that it was "neutral" in regard to religion.[91] If a law is found to be "neutral" in regard to religion, the courts ordinarily find that the law does not violate the Constitution.[92] Any argument that the Arkansas law was religiously neutral because it did not mandate the teaching of creationism, or the teaching of human origins generally, was squarely rejected by the Court's reasoning.[93] The law excluded only discussion of evolution but not discussion of creationism or human origins generally.[94] Therefore only the religiously disfavored view was excluded.[95]

Anyone who believed that the debate over teaching human origins in the public schools would die down after *Edwards* failed to learn from the events following *Epperson*, events that spurred the legal battles that arose under the banner of creation science.[96] As legal battles over creationism begat battles over creation science, creation science would soon beget a warrior offspring that would attempt to adapt to its environment to avoid legal defeats. That offspring is, of course, ID.[97] The move from creationism to creation science had caused a rift among creationists while providing creation scientists with new legal ammunition.[98] After that ammunition misfired, the move toward ID created a firestorm of controversy within which we still dwell.[99]

What is most remarkable, however, is how poorly adapted ID appears to be to succeed in the current legal environment. The movement may be counting on a change in the legal environment and, as discussed in Chapter 7, it may well see one. Yet ID's core is still focused on a supernatural designer, and as you will see in Chapter 2, the one court that considered the question did not view ID all that differently from creation science, despite ID's attempt to avoid naming the designer.[100] ID's other main argument requires that schools "teach the controversy,"[101] either by teaching ID when evolution is taught or calling the validity of evolution into question when evolution is taught.[102] This argument has a focus surprisingly similar to that of the balanced treatment laws struck down in *Edwards* and other cases,[103] but because it does not require that ID actually be taught, it may be more legally plausible. As will be seen, however, plausibility does not equal success, and this argument, too, is likely doomed to fail.

III. Understanding Intelligent Design

In *Creationism's Trojan Horse: The Wedge of Intelligent Design*,[104] Barbara Forrest and Paul R. Gross painstakingly document the history of the ID movement. They suggest that ID was designed, in part, as a strategy to get around the numerous legal defeats that both creationism and creation science endured.[105] In fact, however, this link may be even greater than Forrest and Gross argue. Early ID supporters read the language in *Edwards* and other cases and realized they had to take God out of their approach to get design into the public schools and into scientific discourse more generally.[106] They also realized that they would need to do work that could at least plausibly be called science and that they would need to gain acceptance for this work at least among the general public.[107]

This is why many people perceive ID as being nefarious. It is a marketing strategy designed to gain legal and cultural acceptability. Much of what we see today, including Ben Stein's recent movie,[108] is that strategy in action. Yet as discussed in later chapters, ID is a form of religious apologetics, and as such, it is not so much nefarious or deceitful as it is religious marketing.[109] To understand this, it is essential to understand the nature of religious apologetics and the various perspectives that inspire religious apologists. This section is designed as a basic overview of ID. A more detailed discussion of the ID movement's apologist roots will be left to later chapters.

For now it is important to keep in mind that religious persuasion and proof is the focus of religious apologetics, and at the core, most apologists believe in the truth of their ultimate goal.[110] Thus, while those who trust modern scientific methodology often view ID advocates as charlatans because of the weak scientific arguments they use, ID advocates sincerely mean it when they claim that their position is valid and is a valid alternative to mainstream science.[111] They see mainstream science as biased against their ultimate truth.[112] From the perspective of the religious apologist, the end of serving ultimate truth justifies the means.

Among the originators of the ID movement are two law professors, Phillip Johnson and David K. DeWolf.[113] One might expect that biologists would be the primary originators of what is claimed to be a scientific approach, but when one looks at the early proponents of ID, there were more philosophers, law professors, and social scientists than natural scientists; a number of the natural scientists were not biologists; and none of the biologists were evolutionary biologists.[114]

In fact, it was Phillip Johnson, a law professor, who spurred the movement with the 1991 publication of his book *Darwin on Trial*.[115] Interestingly, the book starts out by discussing *Edwards v. Aguillard*.[116] From there, Johnson moves into an attack on "scientific materialism," which Johnson defines as attempts "to explain all human behavior as the subrational product of unbending chemical, genetic, or environmental forces."[117] His ultimate assault in the book is on evolution through natural selection, commonly referred to as Darwinian evolution.[118] Attacking evolution through natural selection remains an essential part of the ID movement's strategy today.[119]

Johnson's worldview, and the ID worldview generally, appear grounded in several assumptions: (1) there are absolute moral principles by which humans are meant to abide;[120] (2) Darwinian evolution removes the basis for such principles by treating human existence as a series of unguided biological accidents (their characterization, not a necessary or even accurate one);[121] and (3) Darwinianism promotes scientific and natural materialism, that is, the view that unguided natural forces are responsible for everything.[122] Johnson's book was largely ignored outside the ID and Evangelical communities. When the book was noticed by the mainstream scientific community, the so-called scientific claims made in it were quickly discredited.[123]

In 1992, soon after publication of his book, Johnson was joined by other early ID proponents.[124] By 1996, the Center for the Renewal of Science and Culture was founded at the Discovery Institute.[125] At this point, Phillip Johnson and other ID supporters were already working on what has come to be known as the *wedge strategy*.[126] In *Creationism's Trojan Horse*, Forrest and Gross document the evolution and implementation of this strategy.[127]

The so-called Wedge Document was produced by the Discovery Institute in 1998. It is essentially a game plan and marketing strategy for ID.[128] Interestingly, the wedge strategy seems an odd vehicle to support a supposedly scientific approach because it is primarily focused on gaining acceptance for a preconceived notion of human existence, and even its discussion of scientific research is couched in terms of gaining acceptance for ID.[129] There is nary a mention of specific scientific methodologies (as opposed to goals) to be used by ID proponents; nor is there any mention of research that could possibly falsify ID's core assumptions (this is somewhat ironic because, as will be seen, this is exactly what ID proponents accuse evolutionary biologists of failing to do).[130] The wedge strategy exposes the apologist, specifically the Christian apologist, roots of the ID movement.

The Wedge Document was not originally intended for public consumption. It was leaked and then published on the

Web.[131] This has proven problematic for the ID movement because ID proponents have sometimes argued that ID is not an inherently religious or theistic concept in an attempt to distinguish ID from creation science.[132] Of course, given that the roots of ID are in Christian apologetics (and in natural theology, itself a touchstone of apologetics), ID advocates' attempts to deny that the designer is divine are a reasonable response – from a legal and marketing perspective – to the language in *Edwards* prohibiting the teaching as science of religious theories that are not falsifiable. It is an attempt at shielding ID from legal attacks under the establishment clause.[133]

The Wedge Document is something of a hole in this armor, however – a hole that had a significant impact on how the one court to discuss the ID concept in depth viewed ID.[134] None of this would be an issue, of course, if ID proponents did not insist on engaging in the "proof game." Religious thinkers from Thomas Aquinas to Reverend Paley have argued for theistic design,[135] and there are numerous philosophical and religious arguments in favor of God as an intelligent designer. Of course, these religious and philosophical arguments cannot be proven in any scientific manner and cannot be taught as science in public schools (they can be taught in philosophy and comparative religion classes so long as they are taught objectively and not as religious truth).[136]

In fact, neither Aquinas nor Paley saw any reason to argue that design is not a religious concept.[137] These were clearly faith-based observations of the world around them, observations with which many still agree. Both were openly religious apologists and would probably scoff at the ID movement's public denial of the divinity of the designer.

So why seek to prove these ideas scientifically? Two reasons are apparent. First, for religious and social reasons, ID proponents view scientific materialism as dangerous and a necessary, or at least important, component of moral relativism (a point with which many scientists and philosophers would disagree).[138] ID proponents contrast this with supernaturally inspired views of nature and the world and moral absolutism, which they believe is necessary.[139] Second, the law! The courts have been clear that faith-based views on creation might belong in philosophy or comparative religion classes but do not belong in science classes.[140] So to be able to reach the hearts and minds of the nation's youth in public schools and universities to persuade them against scientific materialism, ID must be viewed as scientific and not grounded solely in religion.

A. A basic primer on ID: Intelligent design 101

Two overlapping strategies are generally used by the ID movement: first, exploiting gaps in evolutionary biology and attacking evolutionary biology generally[141] and, second, trying to demonstrate the designer through the complexity of living organisms.[142] The end goal of both these approaches is to overthrow scientific materialism and what ID proponents call methodological *naturalism*.[143] Naturalism, as defined by

ID advocates, is the idea that natural forces explain what we see in the world and in living organisms (this is how science defines naturalism) *and* that the world and the organisms in it came about through purely natural (i.e., no higher power) mechanisms.[144] This latter point is central to the ID argument against naturalism. Methodological naturalism basically comprises the methods scientists use to study the natural world.[145] This includes the scientific method, thus, the ID argument is aimed, at least in part, at the scientific method itself.[146]

Interestingly, the argument based on the exclusion of a higher power, which ID advocates claim to be inherent in methodological naturalism, is a straw man argument. One can accept naturalism and the mechanisms said to support it without denying a higher power. Naturalism certainly includes the first element of the preceding definition, but the second element is not a necessary or even useful part of the definition of naturalism.[147] In fact, biochemist Kenneth Miller wrote extensively about this in his influential book *Finding Darwin's God*.[148] Theistic evolution, the subject of Chapter 4, is based in the belief that science and religion can coexist happily.

It is only because ID proponents enter the scientific proof game that their straw man takes on life. Many people of faith accept what ID proponents call methodological naturalism – think *scientific naturalism* for lack of a better term.[149] *Methodological naturalism* is just a fancy term for the scientific method as applied to natural phenomena and natural processes.[150] Naturalism, methodological or not, is not inherently inconsistent with faith, nor does it preclude the theological notion of God as designer.[151] For people of faith who accept scientific evidence, naturalism may simply suggest that the natural mechanisms observed and documented by scientists are the work of God[152] – the latter point, of course, being beyond scientific proof. This is not a problem until one assumes that (1) naturalism somehow must conflict with faith and that (2) science is the appropriate arena in which to try to prove the existence of the supernatural/divine. ID assumes both these things.[153]

Another central aspect of ID is the tendency to deny that the designer is God.[154] Yet many ID proponents have suggested that God is the designer,[155] and the Wedge Document is explicit about it, so why not just come out and say it? *Edwards v. Aguillard* and other legal cases may be one reason. The need to gain acceptance as a scientific approach may be another. Yet when one reads about the ID movement both from its supporters and opponents, it seems obvious that the designer they have in mind is God.[156] For present purposes, however, I will take ID advocates' suggestion that the designer need not be divine at face value and thus will not refer to the designer as God, except where others have done so.

This, however, creates something of a dilemma when writing on this topic. Constantly referring to the "designer" could get a bit tedious. When I have spoken to general audiences on this topic, I have frequently referred to the designer as "Fred" to emphasize ID proponents' general refusal to openly acknowledge that the designer is God. Another name I have used is "Beatrice." Yet in this book I have settled on a name that should work without choosing a gender for the designer. Thus, throughout the rest of this book the designer will be referred to as "Big D." We will assume that Big D is not necessarily God, even though he, she, or it fits a big part of the job description.

One of the first things many ID supporters will tell you is that evolution is just a theory.[157] The definition of the term *theory* has itself been the subject of volumes,[158] but what ID supporters generally mean by this is something along the lines of "it's just a theory, not fact."[159] Again, this is a straw man argument because calling something a theory to cast doubt on its accuracy only works if you believe that there is an absolute truth that is absolutely provable without the need to build on prior knowledge. Of course, many scientists and philosophers use the term *theory* in ways different from ID supporters.[160] Still, the existence of evolution is no longer really a *theory* in the way ID supporters use that term.[161]

What does this mean? It means that mainstream science has demonstrated that evolution occurred and occurs.[162] The dispute within mainstream science is not over whether evolution is real but rather over specific aspects of evolution such as causal mechanisms for certain types of mutations, what drives certain changes at the genetic level, and how a specific environment might affect organisms when that environment experiences climatic or other major changes.[163] No credible mainstream biologist would argue that evolution, including human evolution, did not happen and is simply unproven; rather, biologists will use scientific means to try to prove or falsify specific hypotheses, and over time, more and more specifics will be understood.[164]

This is how modern science works. No credible scientist, regardless of his or her scientific discipline, will claim, "I have the absolute answer to all the questions surrounding my theory." Scientists try to prove or falsify an underlying hypothesis as well as the questions it raises.[165] Frequently new data begets new questions. The key is that scientists generally try to *prove or falsify* a hypothesis, not just prove it.[166] To proceed otherwise would lead to situations in which, for example, someone believes the solar system is the center of the universe, makes an anecdotal argument to prove this point, and assumes its truth. This, of course, has happened historically,[167] and it is not far off from the approach the ID movement takes toward the ultimate question raised by ID: the existence of Big D.

To say that evolution is just a theory makes as much sense as saying it is just a theory that the Milky Way galaxy is part of a cluster of galaxies. There is plenty of scientific evidence that our galaxy, the Milky Way, is part of a local cluster of galaxies,[168] even if scientists are still working out some of the specifics regarding why galaxies cluster in certain ways and how this clustering affects neighboring galaxies.[169] Thus the galaxy cluster is scientifically demonstrable, but some of the mechanisms and phenomena regarding galaxy clusters are still being worked out through scientific work designed to *prove or falsify* hypotheses about the mechanisms that drive galaxy clustering.[170] To refer to galaxy clusters as "just a theory" because of this would be inane.

Interestingly, ID proponent Guillermo Gonzales has essentially argued that the natural mechanisms astronomers have documented are not natural but rather evidence of Big D.[171] Many of his arguments resemble those discussed later regarding *irreducible complexity* and *complexity by design* only at the cosmological rather than biological scale. Should we now question work by astronomers and physicists on the natural phenomena causing galaxy clusters?

ID has trapped itself in the proof game, and it does not have the tools to get out of the trap. Arguing that evolution is just a theory – even if one takes that argument at face value – does not prove the involvement of Big D, and ID's attempts to demonstrate Big D through science have done little more than rehash long-standing theological arguments cloaked in supposedly scientific jargon.[172] Let us take a closer look at how ID attempts to prove its hypothesis. We cannot look at how ID attempts to *falsify* its hypothesis because unlike mainstream scientists, ID proponents do not ever attempt to falsify the existence of Big D.[173] Big D's existence and role are assumed.[174]

B. Complexity and design: Reinventing Paley's wheel

In 1802, Reverend William Paley published *Natural Theology*.[175] In his book, Paley discussed the concept of the watchmaker God.[176] His book was part of an important and broader movement particularly popular in the 17th and 18th centuries that looked to relate the natural world and religion.[177] Paley, like other theological naturalists, studied the natural world quite seriously and through the lens of how the natural world reflects the divinity of God.[178] Paley's watchmaker analogy can be restated roughly as follows: a person walking through a park comes upon a stone on the ground and in it may see the natural world at work without regard to design. That same person walks through the park again and comes upon a watch. The person, upon observing the watch, is likely to recognize that the watch must have been designed by an intelligent creator, and thus the analogy proceeds to equate complex natural phenomena to the watch and the watchmaker to an intelligent (and divine) creator.[179] Even in Reverend Paley's time, this reasoning was not new. The idea goes at least as far back as Plato's often overlooked dialogue, *Timaeus*.[180] Analogues can be found in the Roman philosopher Cicero's *De Natura Deorum* (*On the Nature of the Gods*)[181] and in Thomas Aquinas's *Summa Theologica*.[182]

Of course, Paley, unlike many ID proponents, did not claim that the concept was new, nor did he attempt to hide its connection to the divine. Reverend Paley was unabashedly a Christian apologist.[183] Thus, he was committed to using nature to prove the truth of Christian teachings.[184] Paley had no reason to hide this, and in fact his work, when viewed as a work of Christian apologetics and natural theology, was impressive for its time.[185] But that, of course, is the point. Reverend Paley would not deny that the designer (watchmaker) is God, and he would not deny that natural theology is theology. Paley wrote decades before Darwin published the *Origin of Species*,[186] but Darwin was clearly influenced by the meticulous observation of nature in Paley's work – Darwin also held the same chambers at Cambridge University that Paley had held,[187] but while Darwin saw the obvious benefits of Paley's detailed observations of nature, he rejected the ultimate assumptions inherent in the work of natural theologians.[188] Darwin was, of course, far more concerned with proof in nature than the metaphysical assumptions inherent in natural theology.[189]

The point is that the watchmaker argument was not new even in the early nineteenth century, although Reverend Paley's explication of it was quite advanced for its time. Yet ID proponents claim to have developed new theories, such as irreducible complexity and specified complexity, that bear a remarkable resemblance to natural theology and creationist arguments.[190] What the ID proponents have done is repackage these old ideas without explicit reference to the divine, sprinkle in some fancy terminology, and apply the watchmaker idea to new contexts, such as biochemistry, to make it all sound more scientific.[191] So let us explore irreducible complexity and specified complexity in light of Paley's work, mainstream science, and the ID movement's frequent denial that Big D is God.

Michael Behe, a biochemist and leading proponent of ID, proposed the concept of irreducible complexity in his well-known ID tome *Darwin's Black Box: The Biochemical Challenge to Evolution*.[192] In it, Behe defines irreducible complexity:

> By irreducibly complex I mean a single system composed of several well-matched, interacting parts that contribute to the basic function, wherein the removal of any one of the parts causes the system to effectively cease functioning. An irreducibly complex system cannot be produced directly (that is, by continuously improving the initial function, which continues to work by the same mechanism) by slight, successive modifications of a precursor system, because any precursor to an irreducibly complex system that is missing a part is by definition nonfunctional.[193]

Behe compares such a system to a mousetrap.[194] He suggests that if one removes a part of a mousetrap the trap will no longer function and analogizes this to an irreducibly complex biological function.[195] He claims that if you remove a part of such a system it would no longer function, and thus such a biological system could not have evolved through natural selection because it needs all its parts to function.[196]

Commentators have demonstrated both the logical and biological fallacy inherent in this argument.[197] First, as scientific philosopher Robert Pennock points out, if you remove a part (or parts) of a mousetrap, it might cease to function as a mousetrap, but what remains will still function for other purposes.[198] Biochemist Kenneth Miller has suggested that you could remove the catch and metal bar and still have a fully functional tie clip or paper clip.[199] Others have pointed out that you could remove the wooden base, but the trap would still work if it were attached to its natural environment: the floor.[200]

Miller also demonstrated that each of Behe's proposed irreducibly complex biological and biochemical systems has been shown to be a product of natural selection.[201] This includes wings,[202] bacterial protein sequences,[203] and the blood-clotting mechanism[204] that was the centerpiece of Behe's argument.[205] Moreover, in recent years, one of Behe's other major examples, the bacterial flagellum – Miller has explained how the flagellum could have evolved[206] – has been shown to have precursors (and living relatives), essentially microscopic poisonous stingers that are closely related to flagella and would have been favored by natural selection to evolve into flagella.[207]

The point is that Behe proves too much with his mousetrap analogy and other examples. If you remove part of a mousetrap, you may no longer have a functioning mousetrap, but you could still have a functioning device.[208] Of course, evolutionary biologists have long known that even in complex organisms, a function may evolve for one purpose but eventually may come to serve another through the process of natural selection.[209] Miller's response to

Behe's mousetrap analogy demonstrates this nicely. In fact, the scientific evidence against irreducible complexity is overwhelming.[210]

Moreover, irreducible complexity itself is just a rehash of Paley's watchmaker, which of course comes from a work of Christian apologetics and natural theology! Behe simply removes explicit reference to God, adds the connection between irreducible complexity and biochemistry, and sprinkles in a lot of fancy scientific terminology that in the end simply describes Reverend Paley's watchmaker in the biochemical context.[211] In his book on the evolution of the ID movement, *Tower of Babel: The Evidence against the New Creationism*,[212] Professor Robert Pennock, a highly regarded philosopher of science, demonstrates the connection between ID theory and theistic naturalism, including the connection between Behe's work and Reverend Paley's.[213]

Behe is not alone, however, in recasting natural theology in supposedly scientific terms. William Dembski, perhaps the most prolific ID proponent and a faculty member at the Southwestern Baptist Theological Seminary, argues for *specified complexity*.[214] What is specified complexity? It is an approach grounded in mathematical and logical assumptions that suggest patterns that are both specified and complex – that is, patterns that demonstrate characteristics evincing intelligence and that cannot easily be explained by chance – are evidence of design by an intelligent actor.[215] Dembski applies this concept to biological functions in the

ID context.[216] Dembski has explained the criteria for specified complexity as follows:

> Whenever we infer design, we must establish three things: *contingency*, *complexity*, and *specification*. Contingency ensures that the object in question is not the result of an automatic and therefore unintelligent process that had no choice in its production. Complexity ensures that the object is not so simple that it can readily be explained by chance. Finally, specification ensures that the object exhibits the type of pattern characteristic of intelligence.[217]

Mathematicians and logicians have demonstrated the numerous assumptions and flaws in this approach,[218] and Dembski's own definition demonstrates the problem. His argument is entirely circular. You must first assume that his three prerequisites to infer design indeed infer design; only then can you look to them to prove his point.[219] Yet mathematicians have demonstrated that each of Dembski's three criteria, and his application of them, involves misstatements and mischaracterizations of data and that, in fact, Dembski's use of *specified complexity* to demonstrate the probability of Big D is impossible because the data he uses are inadequate to prove how the various biological mechanisms he describes came into being.[220]

C. Big D: The filler of gaps

ID proponents often rely on two additional arguments. The first involves the gaps in evolutionary biology (this also relates to Behe and Dembski's work discussed earlier).[221] The second involves the concept of *teaching the controversy*.[222] These will be discussed in turn.

Even if irreducible complexity and specified complexity were junk theories, ID proponents can still fall back on their argument that evolution is just a theory and that it contains numerous gaps. This was discussed a bit at the beginning of this chapter, but we will explore the argument in more detail here. The argument is essentially that Big D can be found in the gaps in evolutionary theory.[223] It is the notion of "Big D and the gaps," or as others have suggested, the "God of the gaps."[224]

As was explained earlier, however, any credible scientific theory about anything remotely complex is likely to have gaps as scientists work out the specifics of the theory.[225] The existence of gaps proves nothing one way or the other. Only someone trying to market a position would suggest that the gaps clearly imply anything other than an area that scientists are still exploring. There is essentially nothing to infer other than a potentially fertile area for research. The existence of evolution has been scientifically demonstrated.[226] The unresolved aspects of evolution revolve around how specific aspects of the evolutionary process unfold. Any gaps demonstrate only that scientists do not have an answer to a specific question. It says nothing about *any* answer to that question.[227]

Thus, filling the gaps with irreducible complexity,[228] specified complexity,[229] the Flying Spaghetti Monster,[230] or aliens from the planet Pretenz says nothing about the validity of evolutionary biology, nor does it demonstrate any evidence of Big D or any other supernatural gap filler. Advocates of a supernatural gap filler would still have to independently prove the validity of their theories. ID proponents do not have the tools to do so, and concepts such as irreducible complexity and specified complexity prove nothing other than the creativity and preconceptions of those who proposed those concepts.[231]

From a religious perspective, one might find ID proponents' arguments regarding gaps to be quite offensive. God is not just the God of the gaps for most monotheistic religious traditions.[232] God is the God of everything.[233] From the perspective of theistic evolution, there is no need for a Big D of the gaps because evolution is a scientific fact, and it is a matter of faith that God used that mechanism.[234] At the other end of the religious spectrum, the Big D of the gaps is also alien to young Earth (and some old Earth) creationists because they believe that God created everything in the manner stated in the Bible.[235]

Why would ID proponents make an argument that is so belittling to Big D? ID proponents are trying to market their approach so that it can be taught or referenced in public school science classes. Thus, they cannot rely openly on the natural theology approach of Christian apologetics, and they need to poke holes in their self-perceived arch nemesis, evolutionary biology.[236] It all comes down to the proof game. If ID did not engage in the proof game, there would be no need for the Big D of the gaps. ID could be taught as a long-standing theological or philosophical approach that directly presumes that God is the designer, and it would be subject to counterarguments from theologians and philosophers rather than scientists. Significantly, ID would be better equipped to engage in such a debate than it is to engage in a scientific debate with scientists. Keeping it a theological or philosophical argument would, of course, exclude it from science classrooms under the establishment clause of the First Amendment to the U.S. Constitution, and that would conflict with one of the major reasons for ID theory: marketing a brand of creationism in science classrooms without acknowledging that it is a form of creationism. It would also limit ID's ability to claim its place as an alternative *scientific* theory to evolution.

Finally, ID proponents rely on an argument focused on teaching the controversy. By this they mean that ID proponents should advocate that teachers, professors, public personalities, and so on, teach about the controversy between ID and mainstream biology.[237] This is a clever rhetorical move. By suggesting that there is a controversy to teach, ID proponents attempt to legitimize ID.[238] Furthermore, by suggesting that the alternative to ID is mainstream evolutionary biology and that the disagreements between the two should be taught, ID proponents are able to place their approach on the same rhetorical playing field as mainstream science.[239] In fact, Ben Stein's movie focuses heavily on this concept.[240]

But, alas, this too is a red herring. It is a brilliant rhetorical move and a wonderful use of smoke and mirrors, but the argument to teach the controversy proves too much. This will be discussed throughout the book, but the following provides an overview of the problems with this argument.

First, the whole notion of teaching the controversy assumes that there is an actual controversy to teach about. From the perspective of mainstream science, there is not.[241] ID is seen by many mainstream scientists as intellectually dishonest, not real science, and so on, but it is not seen as a potential challenger to mainstream science. Chapter 2 will address the notion of scientific paradigms, on which ID proponents sometimes rely to support the argument that there is a controversy.

Can something be a controversy when only one party views it as such? For example, let us say that I believe heavy metal music is the best type of music, and you believe country western is the best. You are quite committed to this position and find my taste in music troubling. I, on the other hand, do not care one way or the other about your taste in music but just wish you would stop annoying me with arguments about why Mel Tillis is superior to Tesla (which, by the way, is a scientific impossibility).[242] Your comments are irrelevant to my musical choices. Is there really a controversy? The controversy is all in your head. I do not care. I will go on listening to heavy metal and continue to ignore your arguments. You may choose no longer to be friends with me because of my refusal to accept your truth, but from my perspective, there is no controversy outside that in your head.

The preceding hypothetical scenario focuses on the question of whether there can be a controversy between parties when only one party perceives and treats the controversy as such. This is similar to the relationship between ID and science. Mainstream scientists do not see ID as a competing scientific theory but rather as an annoying distraction from real science.[243] If they engage with ID proponents, it is generally to show the scientific flaws in ID, not to suggest that there is a valid controversy.[244] Of course, anytime a mainstream scientist engages with ID, the engagement can be used to add further rhetorical fuel to the notion that there is a controversy, even though the scientist sees no *scientific* controversy.

Another example that more directly parallels the implications of the ID movement's teach the controversy approach demonstrates the results of taking that approach at face value. Some UFO advocates believe that humans were placed on Earth by aliens or at least that our biological ancestors were.[245] Significantly, this is different from the idea that life, or its building blocks, may have arrived on Earth from a comet or other extraterrestrial object. The latter theory suggests that the molecular building blocks for life, or maybe even microscopic biological organisms, might have arrived on material that impacted the Earth early in its history.[246] The UFO advocates to which I am referring suggest that advanced civilizations seeded the Earth or placed fully developed organisms here.[247] Certainly, arguments made by ID proponents, especially irreducible complexity and the gaps argument, might be used by these folks, and the UFO advocates might argue that there is a controversy based on their core beliefs and these assertions.

Of course, the key is that they have no direct proof that any advanced alien civilization would be capable of reaching the Earth, let alone that such a civilization seeded the Earth. Should public schools teach this controversy? Is it science? If your answers are "no," how is this any different from ID? This same hypothetical, of course, could be repeated using alchemy, astrology, and any number of other similar concepts. Thankfully, astrologers, UFOlogists, and alchemists do not generally argue that their approaches should be taught in public school science classes or accepted by mainstream science without convincing scientific proof. Whatever their beliefs, they do not market their concepts as *the* alternative to methodological naturalism in the same way ID advocates do.

We will return to the argument for teaching the controversy in later chapters. For now, it is helpful to understand that this may be the ID movement's best argument for legal purposes and from a marketing perspective in the court of public opinion. To create such a controversy, however, ID advocates need to redefine science. That is the subject of the next chapter.

2

Big D and Manufactured Paradigm Shifts

Intelligent design (ID) advocates have argued, although frequently without much philosophical sophistication, that reliance on the current scientific model excludes religious or other nonnaturalistic paradigms from science.[1] If such alternative paradigms are to gain any acceptance, the argument goes, it is essential that they be explored by researchers. This raises the related question of whether such arguments for alternative scientific paradigms must be credited by schools and science departments. The latter question will be addressed in later chapters. For now, it is important to gain an understanding of the implications of the ID movement's use, whether intentional or not, of what are known as relativistic arguments in philosophy. As will be seen, the ID movement is not well equipped to utilize these arguments, and perhaps more important, the work of the progenitors of scientific relativism demonstrates why relativistic arguments do not help the ID movement.

I. Intelligent Design and Modern Science

Inclusion in the realm of science generally requires that one "do science," that is, use the tools and methodology of science. To "do science," one must generally engage in work that allows one's hypothesis to be supported or falsified – that is, that could prove that a scientist's hypothesis might be wrong.[2] Falsifiability is one key element of modern science, an element famously championed by philosopher Karl Popper, who attempted to define the parameters of modern science.[3] Still, arguments can be made that suggest that Popper's definition of science is simply one paradigm for science and that there is no superparadigm that allows one to prove the correctness of a given scientific paradigm.[4] As explained later, this argument can be supported by the work of the scientific historian Thomas Kuhn,[5] but ironically, Kuhn's arguments when taken as a whole work against ID.[6]

From a legal perspective, moreover, the possibility of multiple scientific paradigms raises some interesting questions when merged with free speech concerns and the concept of equal access to government forums and programs. This will be the topic of Chapters 6 and 7. ID proponents' argument that they are denied academic freedom, excluded from scientific debate, and subject to discrimination raises the question of what exactly they are being excluded from.

If ID is not science, excluding it from scientific fora makes perfect sense. For example, we do not expect that interpretive philosophy will be taught in a physics class, funded under a grant designed to support physics research, or published in an academic journal devoted to physics. If ID is science, however, it can be argued that excluding it from government-controlled or -sponsored scientific fora, because it is perceived as religion, raises serious constitutional concerns.

ID proponents make no attempt to falsify their ultimate hypothesis that Big D exists and that Big D designed many complex organisms, nor do they attempt to falsify their arguments that evolution cannot explain much of what is seen in the natural world or directly try to prove Big D's existence through experimentation that can be reproduced, so it is obvious that ID is not science under the traditional definition of that term in the era of modern science.[7] If, however, ID advocates can argue successfully that the traditional definition of modern science is not the only one, and better yet, that it "discriminates" against alternative "scientific" approaches, suddenly ID can try to claim the mantle of free speech protection even in the scientific context.

Of course, this whole argument is question begging because one could make similar arguments about the paradigms for everything. Thus, in the sciences, alchemists could argue for access to the scientific forum, and using a divining rod to find water might have to be included in courses or programs on geology or Earth sciences.

Meanwhile, in the social sciences, every possible conception of social ordering could be given equal credence, regardless of its academic merit or acceptance within the relevant academic discipline (even among the avant garde of that discipline). This will be discussed further subsequently.

II. It Is All Relative, or Is It? Some Basics on Kuhn and Scientific Paradigms

The possibility of multiple scientific paradigms was the subject of Thomas Kuhn's seminal work *The Structure of Scientific Revolutions*.[8] Since then it has been the subject of numerous books and articles. At first glance it may appear that ID proponents could use the notion of multiple scientific paradigms to argue that religiously based (or at least supernaturally based) paradigms should be considered science, despite their failure to use the scientific method to analyze their ultimate conclusions. The scientific method itself would only be a tool of particular paradigms for science under this analysis. Of course, this argument would allow alchemy, UFOlogy, and astrology to be considered science as well. One would hardly expect that chemistry departments would accept alchemy as an appropriate teaching or research field, nor would one expect an astronomy department to credit teaching or research focused on astrology.

Yet even if one accepts the epistemological implications of Kuhn's approach, that is, that any scientific approach is value and preconception laden and that alternative paradigms may be equally plausible,[9] ID gains no leverage. There are three major reasons this is so, and each will be explored in turn. First, while *Scientific Revolutions* has long been regarded as an important work in the philosophy of science, Kuhn himself was primarily a scientific historian and viewed his work as such.[10] As will be seen, whereas Kuhn the philosopher may at first glance seem to give ID proponents some leverage, Kuhn the historian demonstrates why ID is not likely ever to gain any practical traction even in the world of shifting scientific paradigms.[11]

Second, even if one were to accept all the implications of scientific relativism, ID proponents are hardly in a position to take advantage. Notions of absolutism, both moral and material, undergird the ID movement.[12] As will be seen, by accepting any sort of scientific relativism, ID proponents destroy one of the central tenets of their own "theory." Winning that particular battle – which is quite unlikely to begin with – loses them the war. Of course, ID advocates may use relativism as a marketing tool to gain traction for ID. This would be consistent with the modus operandi of religious apologists, but this particular marketing ploy has the potential to backfire right into the core of the ID worldview.

Finally, ID advocates have argued that even in the absence of bold scientific relativism of the sort suggested by Kuhn, there is no way to determine a precise demarcation point between science and quasi-science.[13] Though at first this issue appears to help ID advocates, it lends little philosophical aid to ID and it is even less helpful than the

the first two arguments in the legal context. That is, it is less helpful than nothing.

A. Gaining acceptance in normal science[14]

Kuhn's work demonstrates that ID need not be accepted by the community of credible scientists even if it is an alternative scientific paradigm.[15] The possibility of multiple scientific paradigms would be of no practical use to those who could never gain acceptance for their preferred paradigm among mainstream scientists. This is because there is a distinction between the epistemological arguments (arguments about how we know things) made by relativists about the nature of science, which are primarily descriptive, and the normative question of what may be accepted as science in the community of scientists and why.[16] In fact, though at a superficial level it might be argued that Kuhn's philosophical arguments would support the inclusion in science of paradigms that are not based in traditional scientific approaches, when one reads the historical analysis in his work, it quickly becomes apparent that as a practical matter, the opposite is true.[17]

Even within Kuhn's description of scientific paradigms and revolutions, there is a presumed substantive boundary for what may practically be considered science, even if that boundary may shift.[18] Astrology, ID, and the belief that the Earth is the center of the universe are all precluded from "science" today because they do not use the tools, quantitative analysis, or methodology of science in regard to their ultimate hypothesis.[19] Of course, Astrology, Natural Theology and the belief that the Earth is the center of the Universe were accepted scientific paradigms in the past, but scientific evidence caused these paradigms to no longer be accepted by the practicing scientific community. Yet, at an ethereal level they remain scientific paradigms.

Significantly, even if ID can be called a scientific, as opposed to a theological or philosophical, paradigm, it need not and has not been accepted by the community of scientists.[20] Kuhn specifically explains that not all scientific paradigms will be accepted by the scientific community,[21] that the scientific community does determine what "normal science" is,[22] and that there are specific ways in which a new paradigm might come to be accepted by the scientific community or some subset of it.[23] ID, even if it proclaims itself to be a scientific paradigm, has not gained acceptance among credible scientists or scientific journals[24] and is not part of the discourse of the mainstream sciences.[25]

Still, in arguing for academic freedom, an ID advocate might ask how we know that a given paradigm is *the* paradigm for a given science unless there is some superparadigm that allows us to choose between competing paradigms. As Kuhn points out, there is no such superparadigm.[26] I have used a similar analysis in critiquing the concept of neutrality in the religion clause context.[27] Still, as noted earlier, Kuhn teaches us that there are still criteria for what the current scientific community counts as science,[28] and ID does not meet these criteria.[29]

Kuhn's work at most suggests that a theory like ID may have been an accepted paradigm for science – alchemy was based in an accepted scientific paradigm at one point in history – but its methodology and presumptions are so far out of line with mainstream scientific thought that it cannot create a ripple, let alone a shift, in currently accepted scientific paradigms.[30] The reason for this is that ID advocates are unwilling or unable to question their ultimate hypothesis of an intelligent designer, and they have failed to engage in experiments that could support or contravene evolution through natural selection depending on outcomes.[31] In other words, ID advocates work to disprove evolution through natural selection, and their "experiments" and conclusions never account for the possibility that evolution may provide an answer to whatever hypotheses they explore.

ID works toward a predetermined end to disprove evolution, at least as to more complex life-forms.[32] It never seriously considers the possibility that evolution of complex life-forms through natural selection is accurate despite the immense amount of scientific data supporting evolution through natural selection. Significantly, the ID approach is almost the direct opposite of what Kuhn demonstrated to be the historical means that enabled new paradigms to be accepted in science. Kuhn explained that these new paradigms come to be accepted because they give a better explanation of the phenomena they address,[33] they clear up major problems with extant scientific theories,[34] and/or they use better tools than what were previously available to scientists to explain the relevant phenomena.[35]

Therefore it is quite possible that alternative theories can gain acceptance within a discipline by using the tools of that discipline (as well as interdisciplinary tools) to convincingly make the case for such theories[36] or by better explaining the phenomena being studied through evidence (as opposed to presumption).[37] This is, of course, the way that Kuhn suggests most new paradigms come to be accepted.[38] As the primer on ID in Chapter 1 demonstrates, ID provides no new evidence or tools and does not clarify any major problem with evolutionary theory. Nor has it been accepted – and in fact, it has been squarely rejected – by those doing normal science.

Many ID advocates seem upset about their failure to gain acceptance among credible scientists,[39] but as noted earlier, this is heavily a result of their failure to test their ultimate hypothesis – that there is an intelligent designer – through the scientific method.[40] The failure to do so suggests that ID be explored in the humanities, if at all, where philosophy and religious studies leave ample room to explore such questions.[41] It, however, excludes ID from "normal science," which does not accept or value ID because of its failure to use the scientific method to test its many presumptions.[42] One might object that this argument relies on a clear demarcation point for what may be accepted as science, but as will be seen, the demarcation argument does not help ID proponents either.[43]

B. Absolutely relative?

For the sake of argument, however, let us presume that supernaturally and/or religiously based approaches to natural phenomena are valid scientific paradigms, even if their ultimate hypothesis is simply presumed to be correct and no attempt is made to prove or falsify major tenets of the theory through generally accepted scientific methodologies used by those engaged in what Kuhn would call normal science.[44] What might this mean for ID?

Kuhn's philosophical approach, like that of Quine, implicates relativism.[45] If one accepts that there is no way to choose between paradigms without using value-laden assumptions or preconceptions,[46] and there is no place of value neutrality from which one can gauge which values and preconceptions are correct,[47] the result is that there is no way to know that a given paradigm is inherently more correct than another in a metaphysical sense.[48] This dilemma of epistemology raises fascinating philosophical questions that have been the subject of much debate.[49] Significantly, embracing relativism in any fashion, regardless of its broader philosophical merit,[50] undermines the very foundations of ID. Essentially, if ID advocates use relativist arguments to justify inclusion of ID in science,[51] they prove too much.

A core underpinning of ID, whether acknowledged by ID advocates or not, is that Big D's existence is not open to question. Moreover, ID advocates rail against the moral relativism to which scientific materialism allegedly leads.[52] Taking a relativist position on what counts as science, a field of inquiry that deals with the natural world through experimentation requires the acceptance of the underlying tenets of relativism, that is, that human actions, behaviors, and beliefs are inherently value laden, subject to preconception, and that this means there is no way to pronounce that a given theory or belief is better than another.[53] Once one accepts the underlying tenets of relativism, one cannot pick and choose which positions, ideas, and beliefs are better or correct because there is no value-neutral vantage point from which to make such a judgment.[54] Moreover, there can be no such thing as an absolute truth, so from a relativist perspective, there can be no innate preference between Big D, the nonexistence of God, and the Flying Spaghetti Monster.

If one were to accept the position that scientific paradigms not even remotely accepted by mainstream science are equally valid and that such paradigms can or should be taught in public school (and possibly university) science classes, the practical results would be troubling. Astrology, alchemy, UFOlogy, Cartesian vortex theory, Phlogiston theory, ethers, and many more would be viable because mainstream science would hold no greater place than any of these other approaches since no paradigm could claim supremacy. It would not matter whether the alternative paradigms are focused on natural or supernatural causation of natural phenomena. Michael Behe, a leading ID proponent who is also a biochemist, admitted as much when he testified in the *Kitzmiller* case discussed in greater detail in Chapter 3.[55]

Of course, the work of those who suggest that science disproves God would also be a valid scientific paradigm, even though today it can be argued that when scientists step into such discussions, they leave the territory of science and enter that of philosophy and religion. In the relativist world of nonpreferential scientific paradigms, no such boundaries exist. So the writing of ID advocates like Phillip Johnson and Michael Behe might be taught along with Richard Dawkins's *The God Delusion*[56] and the writings of Daniel Dennett.

If ID advocates hope that introducing a paradigm in which ultimate questions may be included in science will help bring people to accept Big D, they may be sorely disappointed. Many teachers and science departments, if required to include ID at the table of science, might do so only if the works of those who deny the existence of God based on science are also included. I have read Dawkins, Dennett, Behe, and Johnson, and my money would be on Dawkins and Dennett persuading more people than Behe or Johnson, and I write this as a person of faith who has little in common with any of the guests at this particular table, given their general all-or-nothing positions.

Johnson and Behe are mainly persuasive to those who already agree with their position or who know little about science, but in the relativist-driven scenario, people would be exposed to Dawkins along with Johnson. Some might suggest that this plays into the hands of ID advocates who frequently exort us to teach the controversy, but they assume that this will be a scientific controversy by presuming that their religiously driven idea would be put up against naturalistic hard science.[57] Yet the best counterpart to ID work in a realm where anything goes in science would be the atheist take on hard science advocated by Dawkins, Dennett, and others. If ID's supernaturally driven approach is science, especially given the fact that the supernatural is simply deductively applied in ID, then Dawkins and Dennett's conclusions about the nonexistence of the supernatural, which are not currently taught as part of science in public school classrooms, could be introduced to the same children who learn ID in those classrooms. Unlike Johnson and Behe, I doubt Dawkins or Dennett would want their metaphysical conclusions taught as part of a high school science class, but if they were taught along with ID, I suspect we would witness an increase in atheism that is at least as large as the increase in ID adherents. This scenario demonstrates why the God–no God debate has no place in a science classroom.

Significantly, it would not stop there. Similar arguments could be made in every academic discipline until the public schools become a public forum for whatever theory or material teachers or maverick school boards want taught. To deny these alternative approaches in any discipline would be to discriminate against such theories and their proponents.[58]

This would apply even when those theories are religiously grounded.[59]

Of course, as noted earlier, Kuhn himself rejected this result.[60] He drew a distinction between the epistemological reality that non-value-laden baselines for judging reality do not exist and the practical reality that the tools of and participants in normal science ultimately decide what counts as science.[61] To create a paradigm shift, a new paradigm must convince mainstream scientists and generally use at least some of the tools of normal science in a manner that is effective in persuading scientists.[62] Therefore, the relativist position – regardless of its philosophical merit – could not be practically used in a manner that would help ID win the supposed origins controversy. ID is again trapped by its own rigidity and the reality that it is more marketing strategy than science.

Importantly, by using relativism – even as a marketing tool to get ID into the schools – ID advocates betray one of their central concerns: that there are moral absolutes and that naturalism, especially evolution through natural selection, leads to the rejection of these values. Absolutes, whether moral or otherwise, are completely inconsistent with relativism. Therefore, if one accepts relativism regarding the physical world, it is more than inconsistent to reject it in the obviously value-laden context of moral theory; it is a farce of epic proportions.

This point cannot be understated. ID advocates need to expand the definition of science to include in science their assertion of supernatural causation and their major deviations from the scientific method as to their ultimate hypothesis, the existence of Big D. Simply put, they have no vantage to do this from within normal science. To include ID at the scientific table, they must fundamentally change what it means to do science. ID advocates have acknowledged as much.[63]

Many of the advances we take for granted today are a result of the fact that science is a field with methodological norms, norms that exclude ID from the realm of science. The only tool that conceptually allows ID to claim a place within the pantheon of science is the one that allows alternative (currently nonscientific) paradigms to count as science. Whether ID advocates call this tool by its name, the tool is relativism, and it devastates the core of ID – that is, ID's embrace of moral certainty and the certainty of Big D's existence.

Simply put, to seriously make the relativist argument, ID advocates must acknowledge that Big D may not exist; that Big D has no superior place in the universe over any conceivable alternative, whether scientific or not; and that intelligent design has no entitlement to be *the* alternative to evolution – all alternatives have equal validity if they claim to be scientific under any conceivable paradigm for science. Moreover, the implications of this approach to science reverberate into other areas. No moral values can be superior to others, except perhaps those on which all people agree (and even these may raise questions under a truly relativist approach), and no religion can have the "truth": religious values have no greater claim to correctness than other values, which of course have no way to claim correctness in any universal sense. Big D is demoted further from the Big D of the gaps to the Big D of one minute possibility in the vastness of conceptual possibility.

If ID advocates want to honestly and openly make relativist arguments to support their cause, they must acknowledge the implications of what they are doing, unless of course it is purely marketing. If so, I call on ID advocates to acknowledge this in one of two ways: first, acknowledge that you are using relativism only as a marketing tool. This would, of course, destroy its value as a tool that any court or thinking person would take seriously as support for ID. Second, follow through openly on your reasoning and acknowledge that there are no moral absolutes, that Big-D is just one of an unlimited number of alternatives to materialism, that religious values and Big D have no special claim to truth, and that ID has no greater claim to science than UFOlogy, astrology, alchemy and ethers or, for that matter, the Flying Spaghetti Monster.

I doubt any such statement is forthcoming from ID advocates, at least not one that would be openly available to those who financially and publicly support ID, but the gauntlet of relativism has been laid down. None of this is likely to keep ID advocates from using the relativist argument as part of their marketing strategy, but that is the subject of later chapters. For now, it is enough to say that whether the ID movement finds scientific relativism useful in its marketing strategy and claims of victimhood, relativism does not help ID enter the realm of science as that field is understood by those actually engaged in it. Moreover, it explodes the ID movement's assumptions about the possibility of moral correctness – assumptions that drive the ID movement's opposition to evolution.

C. Demarcation

Still another argument from the philosophy of science may seem potentially useful to ID advocates, namely, the notion that there are problems in defining the demarcation point between science and pseudoscience or nonscience.[64] The question of whether there is such a demarcation point between science and nonscience has long been a preoccupation of many philosophers of science.[65] The debate goes back to Aristotle and before, and it still rages today.[66] It is a question of epistemology; that is, is there a way to know what counts as science and what counts as nonscience?[67] Ironically, the argument that there is no clear demarcation point between science and quasi-science is itself an argument often supported by relativist concepts.

There are really only three possibilities in the end. Either there is a way to determine what constitutes science and what does not, there is no way to do so, or there is a way to tell whether a specific idea is science, even if there is no way to draw a bright line between science and quasi-science generally.[68] Philosopher Robert Pennock, writing specifically about ID, referred to this third possibility as *ballpark demarcation* and explained that it is adequate to demonstrate that a specific idea, such as ID, is not science even if it cannot draw a clear line applicable to all possible ideas.[69] This seems a pragmatic solution to the problem, and Pennock explains that ID is not in the scientific ballpark.

Of course, many scientists believe that demarcation is not a problem and that there is no need for a middle ground because there are methods to demarcate between science and nonscience, for example, use of the scientific method to test even core assumptions and hypotheses. Therefore those who have proposed clearer mechanisms for demarcating science from pseudoscience or nonscience argue that science uses the scientific method, whereas nonscience does not,[70] or that science can be defined by its growth or predictive ability.[71] Falsifiability is another factor that some argue distinguishes scientific ideas from nonscience.[72] Whether any or all of these factors can clearly demarcate science from nonscience in all circumstances, it is clear these factors help demarcate what constitutes normal science in the sense that Kuhn suggests; that is, the scientific method and ability to account for and predict real-world phenomena through scientific findings are ways in which practicing scientists can tell if something is science.[73] Again, we see a distinction between real-world practice and understanding of what constitutes science and the metaphysical reality that any definition for science – in fact, any definition for anything – is value laden and subject to existential alternatives.[74]

Yet some ID advocates argue that demarcation is incoherent or impossible.[75] The implication of this argument is that because there is no way to tell science from non-science, ID, and in fact supernatural causation generally, is just as scientifically plausible as natural causation. To put it bluntly, the idea that Big D created all complex life-forms, or for that matter magical gnomes from the planet Zermac created life on Earth, is just as scientific as evolution, even if the former ideas cannot be demonstrated or falsified using the scientific method and the latter can.

But this is only true in the realm of metaphysics and epistemology. A practicing scientist would have no trouble knowing that what he or she is doing is science and that ID or magical gnomes are not. In fact, almost every major scientific body in the United States and many around the world have done just that, as have numerous secular and religiously affiliated university curriculum committees and departments (placing evolution in biology and creationism in religion, if it is taught at all).

In fact, the leading opponent of the notion that demarcation between science and nonscience is possible, Larry Laudan, has admirably demonstrated that in the real world, we can tell the difference between valid empirical claims and gobbledygook regardless of the metaphysical inability to demarcate between science and nonscience.[76] Even as he rejects a clear demarcation point between science and nonscience, Laudan points out that "our focus should be squarely on the empirical and conceptual credentials for claims about the world."[77] In other words, the important question is not whether one can know with certainty that something is science or is not but rather what a given practice or conception can show empirically about the world.[78] The label does not matter as much as the substance.[79] This would not appear to help ID advocates given the discussion elsewhere in this book, yet ID advocates regularly rely on Laudan's rejection of demarcation without mentioning this point, which is particularly relevant in the ID context.[80] Like Kuhn, Laudan is comfortable exploring the metaphysical problems with value-laden definitions of what counts as science in some universal sense while still acknowledging that not all scientific paradigms are equal. Some make better claims about the world and back those claims up with better evidence, and normal science favors such approaches.[81]

Still, even if one argues that the lack of a metaphysical demarcation point somehow favors the position of ID advocates, there remain the reality that current scientific practice and opinion reject ID[82] and the reality that to redefine science to include ID opens the door to unlimited and unprovable assertions about the world coming under the rubric of science.

> Doctor: "I'm sorry that the glowing orb hat of Kraplar did not cure your clogged arteries before the stroke, Mr.

Smithers. There was another promising idea that regulating certain proteins in your body with medication would have helped, but there is only a limited amount of funding to study that given all the other areas demanding scientific grants. The good news is that we have some reason to believe that if you wear the barking noodle of gold (BNOG) as a poultice for a few weeks, you may get feeling back in your right side. There is a lot of funding to support BNOG, and it was quite an impressive approach when I learned it in my first year of medical school."

Mr. Smithers: "How do you know it will work? Has it worked for real in others, or was it just a coincidence that some people who used it got better?"

Doctor: "Well, it is supported by evidence. They don't do double-blind studies in BNOG science, but we do know that a number of people got better."

Law and common sense require that all metaphysical possibilities need not be counted as science and that drawing the line where the vast majority of practicing scientists and scientific organizations draw the line both makes sense and avoids scientific anarchy. Any epistemological victory claimed by ID advocates based on the demarcation argument will prove pyrrhic in the legal arena given the applicable law's focus on what is generally accepted by scientists as the best basis for determining what is science,[83] its labeling of supernaturally guided theories generally as religious and thus potentially violative of the establishment clause if taught in schools,[84] and the potential anarchy that would reign if anything could count as science and school curricula or scientific fora were deemed public fora for private speech. Moreover, many, if not most, scientists and even philosophers of science accept that there is a demarcation point between science and nonscience,[85] or at least a ballpark demarcation point,[86] which would further cause the demarcation argument to be of virtually no help to ID advocates given all the other factors just mentioned. In the hands of ID advocates, the demarcation argument does have one thing going for it, however, it sounds really good to those who know little about the philosophical and scientific issues involved. It is a great marketing argument and talking point.

3 The Law and Intelligent Design

I. The Basics

Intelligent design (ID) raises a number of legal issues. The most obvious is whether it can be taught and/or supported in public school science classrooms or whether doing so violates the establishment clause of the First Amendment to the U.S. Constitution. A related issue is whether the inclusion of disclaimers regarding evolution in science classes or texts violates the establishment clause. Additionally, ID advocates have urged that schools should teach the so-called weaknesses in evolutionary theory and methodological naturalism generally – weaknesses that often only exist in the minds of ID advocates and creationists. Another approach urged by ID advocates is to "teach the controversy." The supposed controversy is between naturalism, especially evolution through natural selection, and supernatural possibilities, including Big D.

Yet the legal debate over design does not end with these more obvious legal questions. Central to the ID movement's marketing strategy, and to its potential success on some of the issues mentioned earlier, is the notion that ID is being excluded from scientific fora that include not only classrooms but government funding and the scientific community more generally. Implicit in this argument is the notion that ID has a right to access these fora, and this raises two separate legal issues.

First, when government activity, such as public school classrooms or government funding, is involved, the key question is what type of government forum is involved. If government allows a variety of viewpoints – mainstream science and others – to be expressed in a classroom or program, the situation is more favorable for including ID. If not, there is no need to allow ID access to the classroom or program. This issue will be referred to as the *equal access* issue, and it is discussed later in this chapter.

Second, ID advocates have claimed that they are the victims of discrimination aimed at them by scientific entities and individuals, and by government entities, including public primary and secondary schools and public universities. These allegations raise basic questions about academic freedom and what happens when someone teaches material or does research that is not credited as part of that person's discipline. As will be seen, the key question here is whether

ID advocates are really being discriminated against, and if so, whether it is discrimination in any sense that raises legal – or even moral – concerns. This issue will be addressed in Chapter 6.

II. The Establishment Clause Question

So far only two cases have directly addressed the constitutionality of ID in the public school context,[1] and only one of those cases analyzed ID in any detail. That court had the benefit of detailed testimony by scientists, religious leaders, and philosophers on both sides of the issue, and the court squarely connected the constitutional issues in the case to the question of whether ID is science, religion, or both.[2] As will be seen, one of the problems with marketing ideas rather than testing them, and with trying to market religious apologetics to an audience that does not already agree with the underlying religious ideals – or at the very least cannot legally assume them to be true without admissible proof – is the risk of being exposed as having more pizzazz than substance.

In *Kitzmiller v. Dover Area School District*,[3] a federal district court in Pennsylvania held that the reading of a statement regarding ID in science classrooms, the purchase and placement of ID texts in a school library, and the conduct of the Dover Area School Board, which supported ID, violated the establishment clause of the First Amendment.[4] The key issue in the case was whether ID is religion or science.[5] This issue was so important because if ID is a religiously grounded concept, including it in science classrooms, even through a mandatory statement at the beginning of a class, would violate the establishment clause.[6] If ID is science, however, there might be an argument that it could be included in such classes despite its religious underpinnings.[7] If ID is neither religion nor science, there is no constitutional issue because if it is not religious, the establishment clause could not be violated[8] – although teaching ID as science would still raise serious educational concerns that could be addressed at the state level.

Thus, the best-case scenario for ID proponents would be a finding that ID is not religious and is science. The best-case scenario for those opposing the school board's ID policy would be a finding that ID is not science and is religiously based. It is important to note here that the school board's lawyers specifically asked that the issue of whether ID is science be addressed by the court.

The court heard testimony from leading philosophers of science,[9] biologists,[10] and ID proponents.[11] After hearing all this testimony and evaluating documentary evidence, such as manuscripts of an ID textbook that was virtually identical to a creation science text, with the only significant difference being the substitution of "intelligent designer" for "God" and "intelligent design" for "creation," the court held that ID is not science and that it is a religiously grounded theory.[12] The court's holding that ID is a religiously based theory and not a scientific theory was central to its reasoning under the establishment clause.[13] The Supreme Court had already held in *Edwards*[14] that religiously based theories of creation (in that case, creation science) could not be taught in public school science classes without running afoul of the establishment clause.[15]

Theologians, scientists, and philosophers of science testified that ID is not a scientific theory and that it is a religious theory.[16] Moreover, the school board's top witnesses had a hard time explaining how ID is science and not religiously grounded:

> Moreover, it is notable that both Professors Behe and Minnich [two leading defense experts] admitted their personal view is that the designer is God and Professor Minnich testified that he understands many leading advocates of ID to believe the designer to be God.
>
> Although proponents of the ID [movement] occasionally suggest that the designer could be a space alien or a time-traveling cell biologist, no serious alternative to God as the designer has been proposed by members of the ID [movement], including Defendants' expert witnesses. In fact, an explicit concession that the intelligent designer works outside the laws of nature and science and a direct reference to religion is *Pandas*' [the leading ID textbook] rhetorical statement, "what kind of intelligent agent was it [the designer]" and answer: "On its own science cannot answer this question. It must leave it to religion and philosophy."[17]

Once the *Kitzmiller* court determined that ID is not science, and that it is religion, the outcome that the school board policies violated the establishment clause was unavoidable.[18] In addition to holding that ID is not science and is religious based on the expert testimony and other evidence presented at trial, the court addressed the remarkable behavior by some Dover Area School Board members. Some school board members had threatened teachers,[19] burned an evolution mural that was made by a student and hung in a classroom,[20] laundered the purchase of ID books for the school library through a local church,[21] made brazenly sectarian statements in their official capacities,[22] engaged in sectarian attacks on board members and members of the public who disagreed with them,[23] and lied on the stand and in depositions.[24]

Still, because the court determined that ID is not science and is religiously grounded, all this bad behavior by certain board members was simply icing on the evidentiary cake. Even without this very "un-Christian" behavior, the policy would have been unconstitutional under the prevailing legal tests applied in establishment clause cases.[25] This is important because the court itself did not rely on the behavior of school board members as its primary reason for finding the policy unconstitutional. It relied on its holding that ID is not science and is religion.[26] The ID community itself has acknowledged this, while attacking Judge Jones's opinion for reaching this issue.[27]

To understand the holding in *Kitzmiller*, it is essential to understand the legal tests that courts apply to determine whether government action violates the establishment clause. Thus, a short discussion of the relevant law follows, and after that discussion, the holding in *Kitzmiller* will be addressed in greater detail.

A. The establishment clause

Chapter 1 provided an overview of the two cases involving evolution decided by the U.S. Supreme Court. Both those cases, *Epperson v. Arkansas*[28] and *Edwards v. Aguillard*,[29] were decided under the establishment clause. The *Kitzmiller* court relied heavily on these decisions, especially *Edwards*,[30] but these decisions themselves are part of a much broader body of law that also influenced the *Kitzmiller* court and will undoubtedly continue to influence courts that address ID in the public school context.[31] As will be seen, however, this body of law is anything but straightforward.

The establishment clause of the First Amendment to the U.S. Constitution reads, "Congress shall make no law respecting an Establishment of Religion."[32] The clause became binding on state and local governments in 1947 in the famous case *Everson v. Bd. of Education*.[33] The process through which the establishment clause, and the First Amendment generally, became binding on state and local governments is known as *incorporation*. The incorporation doctrine recognizes that the 14th Amendment represented a shift in state versus federal power[34] – a shift that some argue arose in part because the states may have become as great or even a greater threat to fundamental rights than the federal government.[35] As a result, the Supreme Court began using the due process clause of the 14th Amendment to bind state governments not to harm rights deemed fundamental by the Court.[36]

Many of these fundamental rights are found in the Bill of Rights, and between the 1920s and 1947, the entire First Amendment was incorporated. This is important because many of the disputes that arise in ID situations involve state and local government action. State constitutions often contain provisions similar to the establishment clause, and they too can be relevant in these disputes.[37]

The establishment clause has been characterized – and often mischaracterized – as standing for a variety of sometimes conflicting principles. Thus some argue that it stands for strict separation between church and state,[38] while others argue that it allows government support for religion generally or even monotheism or Christianity specifically.[39] This debate is beyond the scope of this book. I have written extensively about these questions elsewhere.[40] For purposes of exploring the legal questions relevant to ID under the establishment clause, the key is understanding the legal tests and doctrines that apply in such situations.

The legal doctrines applicable to the establishment clause have long been confused and conflicted, but there is some good news for readers of this book. Thankfully – for the reader's purposes at least – there are two areas where establishment clause doctrine has been relatively clear (or at least has led to consistent results). These two areas are government-sponsored or -endorsed religious activity in public schools and equal access for private speech in fora the government has opened for such speech.[41] Significantly, these are the two areas most relevant under the establishment clause in the ID context.[42] The first question will be explored in this section and the other in Sections V.A. and B of this chapter.

When the Supreme Court addressed the constitutionality of teaching creation science under the so-called balanced treatment act in *Edwards v. Aguillard*,[43] the Court applied two legal tests: the *Lemon* test and the *endorsement* test.[44] In a number of cases the Court has treated these tests as interrelated when focusing on the purpose or effects of a given government action.[45] Moreover, since *Edwards* was decided, a third test has evolved in the context of religious activity in the public schools. That test is called the *coercion* or *indirect coercion* test.[46] I will use the latter term. In a recent case involving prayer before football games, the Court applied all three tests.[47]

The *Kitzmiller* court applied only the *Lemon* and endorsement tests,[48] but as will be explained, the indirect coercion test is also potentially implicated if ID is religious, especially if it were to be taught in science classes. This is because it involves subjecting dissenting students (those who do not believe in whatever religious idea the government is promoting) to religious activities the dissenting students have no real option to avoid without feeling coerced.[49] The preceding is just a basic overview of the indirect coercion test, and the subsequent discussion will provide more detail.

1. The *Lemon* test. The *Lemon* test has been criticized for a number of reasons,[50] but it is still good law, at least in some contexts.[51] Religious activity in the public schools is one of those contexts.[52] The test has three prongs. First, government must have a secular purpose for whatever actions or laws are being questioned.[53] In this context, schools would have to have a secular purpose for teaching ID and/or disclaiming evolution.

Second, the primary effect of the law or other government action must neither advance nor inhibit religion.[54] In the present context, that means that the primary effect of teaching ID or disclaiming evolution could not be the advancement of religion. ID advocates might argue that teaching only evolution has the primary effect of inhibiting religion by privileging materialism and naturalism over potential religious explanations. That argument is addressed elsewhere in this book, but it is worth noting here that the argument proves too much. If the reason schools would have to stop teaching evolution is that teaching it inhibits religion, and evolution is a secular theory as the court has held (and this book demonstrates), failing to teach it because it offends specific religions or belief in the supernatural would be to favor religion.[55] This is precisely what the *Epperson* court held violated the establishment clause.[56]

The third element of the *Lemon* test is that there cannot be excessive entanglement between government and religion.[57] Traditionally, this element has had two facets. First is what is known as *institutional entanglement*: concern over government meddling with or overseeing religion, or government and religious entities becoming too intertwined institutionally.[58] The second facet is *divisiveness entanglement*, which occurs when government support of or interaction with religion or religious entities creates divisiveness in the community along religious lines.[59] This latter facet has been eliminated from the test in the aid and funding context but appears to continue to apply in other contexts.[60]

The *Lemon* test is an "or" test, meaning that if any of the three elements are violated, the law or government action is unconstitutional.[61] One reason for the *Lemon* test was to consider the facts on the ground by making sure that government does not substantially further religious ends.[62] Yet the test was not always successful in doing this, and at least in the context of government funding and aid, the *Lemon* test was used to support some absurd results.[63] This eventually led to a modification of the *Lemon* test in the funding and aid context.

In *Agostini v. Felton*, the Supreme Court rolled the entanglement prong of *Lemon* into the effects prong and dramatically limited the importance of the divisiveness factor.[64] The Court also recharacterized the effects test in the aid context.[65] Whether government aid has the effect of advancing religion is now determined by analyzing three elements: (1) Does the aid result in "governmental indoctrination" of religion? (2) Does it "define its recipients by reference to religion"? (3) Does it "create an excessive entanglement" between government and religion?[66] The facial neutrality of the aid program and the mechanism through which the aid is distributed are central to the first two questions.[67] In later cases, such as *Zelman v. Simmons-Harris*,[68] the Court strayed even further from *Lemon* in the aid and funding contexts.[69]

The legal concerns surrounding ID, however, are not generally connected to the funding and aid cases; rather, they closely parallel the school prayer and religious symbolism cases. This is significant because in the Court's most recent school prayer case, *Santa Fe v. Doe*,[70] the Court suggests that all three prongs of the *Lemon* test are applicable when there is a facial challenge to a government policy promoting religion in the schools (although entanglement has not generally been an issue in these cases).[71] This is also reflected in the *Kitzmiller* opinion.[72]

2. The endorsement test. The endorsement was first introduced by Justice Sandra Day O'Connor in her concurring opinion in *Lynch v. Donnelly*,[73] a case upholding the display of a nativity scene by the city of Pawtucket, Rhode Island. The nativity scene was erected as part of a Christmas display in a public park. The display included a number of plastic reindeer and other assorted items. The *Lynch* decision itself has been criticized by many, including this author.[74] The *Lynch* decision, and the criticism of it, are beyond the focus of this book, but the endorsement test that arose in Justice O'Connor's concurrence is quite relevant to legal disputes over ID, and before that creation science.[75]

The endorsement concept was applied in the context of creation science in *Edwards* and in the context of ID in *Kitzmiller*.[76]

Justice O'Connor's concurring opinion in *Lynch* may have given birth to the endorsement test, but many have questioned her application of that test in *Lynch* because she found that the nativity scene did not endorse religion.[77] Still, the test itself has become highly influential, especially in cases involving government-supported or -endorsed religion in the public schools (and in cases involving government display of religious symbols).[78] Justice O'Connor wrote,

> The Establishment Clause prohibits government from making adherence to a religion relevant in any way to a person's standing in the political community. . . . Endorsement sends a message to nonadherents that they are outsiders, not full members of the political community, and an accompanying message to adherents that they are insiders, favored members of the political community. Disapproval sends the opposite message.
>
> Our prior cases have used the three-part test articulated in *Lemon*, as a guide to detecting these two forms of unconstitutional government action. It has never been entirely clear, however, how the three parts of the test relate to the principles enshrined in the Establishment Clause. Focusing on institutional entanglement and on endorsement or disapproval of religion clarifies the *Lemon* test as an analytical device. . . .

The purpose prong of the *Lemon* test asks whether government's actual purpose is to endorse or disapprove of religion. The effect prong asks whether, irrespective of government's actual purpose, the practice under review in fact conveys a message of endorsement or disapproval. An affirmative answer to either question should render the challenged practice invalid. . . .

Focusing on the evil of government endorsement or disapproval of religion makes clear that the effect prong of the *Lemon* test is properly interpreted not to require invalidation of a government practice merely because it in fact causes, even as a primary effect, advancement or inhibition of religion. . . . What is crucial is that a government practice not have the effect of communicating a message of government endorsement or disapproval of religion. It is only practices having that effect, whether intentionally or unintentionally, that make religion relevant, in reality or public perception, to status in the political community. . . .

Every government practice must be judged in its unique circumstances to determine whether it constitutes an endorsement or disapproval of religion. In making that determination, courts must keep in mind both the fundamental place held by the Establishment Clause in our constitutional scheme and the myriad, subtle ways in which Establishment Clause values can be eroded. Government practices that purport to celebrate or acknowledge events with religious significance must be subjected to careful judicial scrutiny.[79]

Justice O'Connor characterized the endorsement test as a clarification of the first two prongs of the *Lemon* test, and the Court sometimes treats it as such today.[80] The Court has also referred to the endorsement test as a separate test, and many lower courts treat it as such even though the analysis overlaps, as we will see in *Kitzmiller*.[81] Significantly, by the time *Edwards* was decided in 1987, the substance of the endorsement test was accepted by a majority of the Court, either as a gloss on the *Lemon* test or, in later cases, as a potentially separate test.[82]

The perspective from which courts are to determine whether a given government activity endorses religion is that of the reasonable observer.[83] The reasonable observer is presumed to be aware of the relevant history and context of the government action in question.[84] Still, there has been debate over whether the reasonable observer is a general reasonable observer or a dissenter or member of a religious out-group in the relevant situation.[85] It is important to note that context is significant both under the endorsement test and the *Lemon* test. The context can be physical, as in the case of religious displays, and also vocal, that is, focusing on the message being sent by the government action. Whether a government action endorses religion is determined by assessing whether a reasonable observer acquainted with the history and context of the government action would perceive that government is creating political insiders and outsiders along religious lines.[86]

Edwards v. Aguillard[87] was decided two years after a majority of the Court first hinted that it would consider endorsement at least as a gloss on the *Lemon* test.[88] In *Edwards*, a majority of the Court applied endorsement of religion as a gloss on the *Lemon* test.[89] As was explained in Chapter 1, the *Edwards* opinion focused on the purpose prong of the *Lemon* test, and the Court held that the Louisiana Balanced Treatment for Creation-Science and Evolution-Science in Public School Instruction Act endorsed religion because it lacked a secular purpose.[90] Creation science was held to be a religious theory, and requiring it to be taught when evolution is taught endorsed religion (as did the provisions of the act that favored creation science by setting aside support for materials).[91] As was explained in Chapter 1, none of the state's asserted reasons was found adequate to survive the purpose-endorsement analysis.[92] As *Kitzmiller*[93] and the recent football game prayer case *Santa Fe Independent School District v. Doe*[94] demonstrate, this result is equally supportable if endorsement is applied as a separate test or approach.

3. The indirect coercion test. Justice Kennedy introduced the indirect coercion test in his opinion for the Court in *Lee v. Weisman*.[95] First, this test is not really applicable in the funding and aid context, but it is potentially applicable in the ID context because it is most relevant where government speaks religiously.[96] That is, in cases involving religion in the public schools and possibly religious symbolism cases,[97] to date, a majority of the Court has not applied it in the latter context.[98] Even in the situations where the test has been applied by a majority of the Court (two cases involving school prayer), it has been used more as a floor below which government cannot go.[99] Therefore many government policies or actions that do not violate the indirect coercion test may still violate the establishment clause.[100]

In *Lee*, a case involving graduation prayer, the Court set forth the core of the test in the following language:

> These dominant facts mark and control the confines of our decision: State officials direct the performance of a formal religious exercise at promotional and graduation ceremonies for secondary schools. Even for those students who object to the religious exercise, their attendance and participation in the state-sponsored religious activity are in a fair and real sense obligatory, though the school district does not require attendance as a condition for receipt of the diploma.[101]

Therefore the dominant factors in the indirect coercion test appear to be that government direct a religious activity and that students be obliged to attend either directly or implicitly.[102] Moreover, the Court held that this sense of obligation can arise as a result of peer pressure and psychological coercion.[103] Therefore, if the other factors are met, even if a school does not require attendance at the religious activity, the indirect coercion test may be violated.[104] This latter point is consistent with earlier school prayer cases that held that allowing students to leave the classroom while prayer is conducted does not validate the otherwise unconstitutional practice of organized school prayer.[105]

This test could easily be applied in the ID context if ID were actually taught in a public school science class. If a court deems ID to be religious – and as this book demonstrates, that is both likely and warranted – students would face a great deal of school pressure and peer pressure to stay in class (assuming there is an opt-out provision). The government would be conducting the class, so there is no question about state action. Admittedly, the situation is even easier under the *Lemon* and endorsement tests, but teaching ID in science would also fail the indirect coercion test.

The only argument to the contrary might be that teaching ID is not a "formal religious exercise," but to the extent that term is a limitation on the scope of the test when the government speaks religiously, teaching a religious idea as science in a science class would be at least the religious equivalent of a short nonsectarian invocation and benediction, which was the "formal religious exercise" at issue in *Lee*, or a prayer before a football game, which was the "exercise" involved in *Santa Fe*.[106] It is not clear that a religious exercise in the sense of prayer is necessary under the indirect coercion test,[107] but teaching a religious idea as science in science class would seem to fit under any sensible interpretation of the language and application of the indirect coercion test. Moreover, though the coercion in *Lee* was unconstitutional because graduation is an event of singular importance and therefore students would feel compelled to attend or miss out on such an important event,[108] coercion in the classroom setting, as the Court has long recognized, is greater and more pervasive.[109]

4. General establishment clause principles. The notion that there is tension between the broad principles traditionally used in establishment clause cases and the results in those cases is not new.[110] Moreover, the relationship between broad principles and the tests used under the establishment clause has been well explored. Yet without reinventing the wheel, it is useful to explore the principles often used in establishment clause cases because those principles interact with the tests that the Court has used; although as this author and others have explained, that interaction is anything but consistent.

The concept of neutrality has always been lurking in religion clause jurisprudence,[111] and as I have written elsewhere, that concept is substantively empty in any of its forms.[112] Neutrality has been used by the Court in several – often mutually exclusive – ways. Thus the Court has treated strict separation as neutral, a rough balance between separation and accommodation of religion as neutral, and most recently, pronounced as neutral a form of extreme formalism that claims the mantle of neutrality while leading to anything but neutral results.[113] As I have suggested, neutrality may be impossible under the establishment clause, but the Court has often used it as a – or even the – underlying principle for its application of a given test under the establishment clause.[114]

Other principles, such as separation, accommodation, equality, and liberty, have also played an important part in establishment clause cases.[115] Writing an overview of these principles is hard to do without either spending hundreds of pages reinventing the wheel or oversimplifying the concepts. I fear that I may do the latter to an extent in this section, but given the focus of this book, it is helpful to at least have a basic notion of these various principles that lurk in the background, and sometimes the foreground, of establishment clause jurisprudence.

Separation was set forth as an establishment clause principle in *Everson v. Board of Education*,[116] a case involving funding for parochial school students to ride public buses in northern New Jersey and the first case to incorporate the establishment clause and make it applicable to state and local governments. The Court held the following:

> The "establishment of religion" clause of the First Amendment means at least this: Neither a state nor the Federal Government can set up a church. Neither can pass laws which aid one religion, aid all religions, or prefer one religion over another. Neither can force nor influence a person to go to or to remain away from church against his will or force him to profess a belief or disbelief in any religion.

> No person can be punished for entertaining or professing religious beliefs or disbeliefs, for church attendance or non-attendance. No tax in any amount, large or small, can be levied to support any religious activities or institutions, whatever they may be called, or whatever form they may adopt to teach or practice religion. Neither a state nor the Federal Government can, openly or secretly, participate in the affairs of any religious organizations or groups and vice versa. In the words of Jefferson, the clause against establishment of religion by law was intended to erect "a wall of separation between Church and State."[117]

Both the majority opinion and the dissenting opinions agreed that the preceding statement reflected the proper approach to religion clause questions.[118] The disagreement between the majority and dissenters was over the application of that approach. The majority upheld the busing program,[119] whereas the dissent believed that the program breached the wall of separation and should have been found unconstitutional.[120]

As a matter of history, the Court almost certainly overstated its case because it relied heavily on an obscure letter written by Jefferson,[121] but the notion of separation was reflected elsewhere in the framers' writings, including other writings by Jefferson.[122] Still, the historical argument for strict separationism is rather weak when one considers all the variables involved in gleaning the intent of the many framers and those at the state ratifying conventions as well as the interpretations and practices of early government entities in the United States.[123]

Yet for the same reasons the case is not any stronger for the competing theories as a historical matter.[124] If we look to the broad intent of the framers and interpret historical practices and principles in light of today's diverse society and massive government, the argument for separation is stronger.[125] This has been called *soft-originalism* in other contexts.[126] I have not, and will not here, take sides in this historical debate – but do note my consistent skepticism about reliance on original intent in the establishment clause context by either side in the debate.[127] Both separationists and those opposing separation have vastly overstated the historical case for their own side, but as we will see, the separationists – at least those who are not strict separationists – may have the stronger argument, at least in some contexts. The strength of that argument, however, does not lie in the intent of framers.

Scholars and the Court have long recognized that "strict separation" is impossible because, at least at the margins, there is bound to be some interaction between religion and government.[128] Strict separation would amount to establishing a purely secular state, where secularism is at least implicitly encouraged and favored and religion banished from the public square and public life.[129] Moreover, it would be impossible, or at the very least highly impractical, to maintain. Another possibility is to use separation as a guiding principle in some contexts but not others.[130] Thus separationist concepts might be used in the school prayer context, the public school curriculum context, and perhaps the direct aid context but not in the equal access or some other contexts. Separation could possibly be used in all contexts, but only when balanced against the other principles undergirding the establishment clause so that it plays a role in defining and applying the legal tests but not the sole role.

Depending on which baseline one picks for separation, it could function as a narrow test in given contexts, a broad principle that urges as much separation between the government and religion as possible, or somewhere in between. Separation is less problematic than neutrality because some degree of separation may be achieved, but separation by itself is problematic at the practical level because one must still choose where and how to implement it. Short of a Draconian absolute separation, which is hard to implement and troubling from a policy perspective, separation is a malleable concept that may function best if implemented as a narrow principle.

Like separation, accommodation can arguably function both at the level of a broad principle and as a narrow principle, or as a facet of a doctrinal test.[131] Accommodationism centers on a governmental structure that accommodates religion. Accommodationist arguments are most common under the free exercise clause.[132] In that context, accommodationism would support exemptions from laws of general applicability.[133] However, accommodationism can also be used in the establishment clause context.[134] As I have written elsewhere, this has often led to results that are troubling both to the nonreligious and the devoutly religious because government generally accommodates religion in a manner that reflects the faith of the majority in a given area but does so in a highly watered-down fashion.[135]

When the Court has relied on accommodationist principles, as it arguably did in *Lynch v. Donnelly*,[136] the results can be hard to justify if one understands much about the nature of religion, in-group/out-group dynamics, or religious symbolism.[137] In *Lynch*, Christmas was declared to be essentially a general or commercial holiday with religious roots, at least to the extent that it is "celebrated" in public life,[138] and the religious impact of a nativity scene – the depiction of the birth of Jesus – while acknowledged, was deemed sufficiently minimized because it was included with secular symbols of the holiday season.[139] The reasoning and results in these cases have been decried by many scholars, both secular and religious, and I need not rehash the rich arguments here. The salient points are that these holdings trivialize a sacred holiday for devout Christians by trying to accommodate public recognition of it without crossing the line into government support of a particular religion[140] and at the same time insult non-Christians by suggesting that Christ's Mass is somehow their holiday too, even if they do not celebrate it.[141] A narrower concept of accommodationism, such as that reflected in equal access cases – cases upholding the right of religious groups or individuals to access government fora that have been opened to private speech more generally – makes a great deal of sense, but as a broader principle, accommodationism has been problematic. Moreover, accommodationism has not been a driving force in cases involving religious activities in the public schools, the cases most relevant in the ID context.

The concept of religious liberty has rarely been invoked by itself in establishment clause cases. It has generally been connected to one of the other principles mentioned in this short section. Courts using religious liberty as a guiding principle must confront its underlying epistemic claim – that is, that there is some way of knowing what religious liberty is. Yet every school of thought that has addressed the religion clauses claims to be promoting religious liberty at some level, and some view their approach as synonymous with religious liberty.[142] Thus religious liberty must either be tied to some baseline or viewed simply as an aspiration to be fulfilled by the doctrine or theory *du jour*. It is a principle that has more potential substance in the free exercise clause context. Under the establishment clause, however, religious liberty is more a platitude than a principle.

Like religious liberty, an approach to religion clause analysis grounded in the quest for equality sounds good. If it could be delivered, all religions would be treated equally, and religion would be treated equally with nonreligion. Judges and scholars who have advocated an equality-based approach to the religion clauses are not naive enough to think that such a state of perfect equality could exist,[143] just as those who have advocated a neutrality-based approach are not naive enough to think that perfect and incontestable neutrality could exist.[144] Yet the Court has used equality principles inconsistently under the establishment clause.[145] Do we measure equality by government purpose? By the facial equality of government action? By the effects of government action? Is treating similarly situated groups the same equality, even if doing so has a disparate impact based on social factors? Is treating differently situated groups the same equality? Existing precedent gives no clear answer to these questions.

The important point here is that equality, like religious liberty, can function as a broad amorphous principle that is never clearly definable or reachable, but it cannot do the work of answering questions in a variety of contexts without the help of some other principle.[146] The situation may be different under the free exercise clause, but this book is concerned only with the ID context, so the establishment clause is the primary focus. One potentially strong use of the equality principle under the establishment clause is consideration of the results of even facially neutral government actions to determine whether those actions give substantial benefits to some religions over others or to religion over nonreligion, which is consistent with the *Lemon* and endorsement tests, but the Court has not been consistent in expressing such ideas even when applying those tests.

Though the previously mentioned principles – and a few others – undergird the Court's application of the legal tests in some contexts, the tests themselves are central to the practical meaning of the religion clauses, even if that meaning has become quite confused as a result of their application. Thus whether a court uses the *Lemon* test,[147] the endorsement test,[148] the coercion test,[149] or some other approach[150] can have a significant impact on the meaning of the establishment clause. The *Kitzmiller* decision reflects aspects of all five of the principles discussed earlier, and this is reflected not only in its application of the legal tests but also in its limitation of the decision to the science class context.[151]

B. The legal holding in Kitzmiller

In *Kitzmiller*, the court applied the endorsement and *Lemon* tests to the school board's ID policy.[152] Judge Jones held that the disclaimer and the other events surrounding the disclaimer (including the acquisition of ID textbooks for the school library) violated the endorsement test[153] and the purpose and effects prongs of the *Lemon* test.[154] Thus, the school board policy violated the establishment clause.[155] Because the court found that ID is not science, and because overwhelming evidence proved that it is religiously grounded, it held that the school board policy violated both the purpose and effects elements of the endorsement test.[156]

The evidence demonstrated that the purpose of implementing the ID policy was to endorse the majority school board members' religious views[157] and that there is no secular purpose that would support their asserting that ID is science.[158] Therefore the policy would make a reasonable observer familiar with the history of the policy feel that the board was creating political and religious insiders and outsiders based on religious views.[159] The board argued that the purpose of the policy was to promote critical thinking skills and improve science education.[160] The court noted that exposure to different ideas and values might support teaching ID in comparative religion or philosophy classes,[161] and perhaps that might serve the goal of critical thinking – as would the introduction of other new ideas in appropriate contexts – but because the court held that ID is not a scientific theory, there was no secular purpose for endorsing it in science classes.[162]

The board fared no better when the court analyzed the effects of the policy under the endorsement test. The court held that because ID is religious and not science, the effect of the disclaimer and book purchases was to endorse religion.[163] Thus, because when the policy was implemented, the disclaimer was read in classes and ID books were added to the library in a well-advertised manner, a reasonable observer would believe that such actions had the effect of creating political and religious insiders and outsiders.[164] There was substantial evidence to back up the notion that indeed this is exactly what happened in Dover when the policy was being debated and after it was passed and implemented.[165] This same analysis essentially applied to the *Lemon* effects test as well.[166] The court used the same reasoning to hold that the primary effect of the Dover policy was to promote the religious idea of ID.[167]

The court's discussion of the school board's purpose under the *Lemon* test went beyond the discussion of purpose under the endorsement test.[168] The court specifically addressed the behavior and statements of particular school board members to demonstrate that the board policy had no secular purpose and was primarily designed to promote the religious views of certain board members.[169] It is important to note, however, that even without such behavior, the court's holdings that ID is not science, and is religious, were enough to cause the Dover School Policy to violate the endorsement test and the effects test under *Lemon*.[170]

Of course, the decision of a federal district court does not have the precedential value of a Supreme Court or appellate court decision, but the *Kitzmiller* court's careful analysis of the science-religion issue will likely be followed by many courts because it is the first decision directly addressing the issue in the ID context and because so many leading figures representing both sides of the issue testified at trial.[171] As mentioned earlier, however, the school board's behavior in *Kitzmiller* was so brazen that courts might use that behavior to distinguish *Kitzmiller*. Yet the Dover School Board's behavior was not the primary focus of the court's analysis on the science-religion issue.[172] It is that issue that was key to the ultimate outcome in the case and that is key to the analysis in this chapter.[173] Therefore, even without the behavior of school board members, the case would most likely have come out the same way.[174]

C. Selman v. Cobb County School District

The only other case to address ID, albeit indirectly, is *Selman v. Cobb County School District*.[175] The opinion of the trial court addressed the constitutionality of a disclaimer placed in the front of a science textbook, which declared that evolution "is a theory, not a fact, regarding the origin of living things."[176] As you learned in earlier chapters, this use of the term *theory* is inaccurate, and of course, evolution is not a theory "regarding the origin of living things." Evolution does not address the origins of life itself but rather change in species through natural selection.[177] The teaching of ID was not an issue in the case, and ID was not specifically mentioned in the disclaimer. The disclaimer was, however, consistent with the ID movement's approach to "teach the controversy" and attempts to discredit evolution as "just a theory" (read *hypothesis*).

The trial court held that the school board did not act with a purpose to promote religion, and indeed the board claimed to have several highly plausible secular purposes, not the least of which was appeasing constituents.[178] Instead the trial court found the disclaimer unconstitutional based on the *Lemon* effects prong, as applied through the endorsement test, and the *Lemon* entanglement prong. The reasonable observer, according to the trial court, would have understood the religious motivations of the constituents that influenced the board's decision,[179] the scientific problems with the assertions in the disclaimer,[180] and would perceive that the sticker is an attempt to undermine evolution to be consistent with the religious views of some members of the community[181] and that the disclaimer policy entangled the district with religion.[182] Obviously the trial court considered the characterization of evolution in the sticker as religiously motivated and such attacks on evolution to be religious.[183]

The trial court opinion, however, has no precedential value. The decision was vacated by the Court of appeals because of serious concerns about the record on appeal.[184] For those unfamiliar with this concept, the record on appeal includes the evidence and legal filings that arose in the trial court. Because the trial court based its holding on the supposed influence on the school board exerted by religious constituents, and because that evidence was absent from the record (along with other highly relevant evidence), the Court of Appeals vacated the decision and sent it back down to the trial court to develop a better record upon which to make a decision.[185] This is why the trial court opinion no longer stands as good law, and it calls into question the findings on which the trial court relied.

As the Court of Appeals held in *Selman*, and the *Kitzmiller* court acknowledged, establishment clause questions are highly fact sensitive.[186] This means, of course, that without an adequate factual record, an appellate court cannot decide the case on appeal.[187] No subsequent opinion has been issued deciding the case.

III. "Teach the Controversy" and the Law

Chapter 1 addressed the ID movement's "teach the controversy" argument. This section will not rehash that discussion except where highly relevant. Rather this section is focused solely on the legal implications of this argument in the public schools; that is, what would happen legally if indeed a district decided to teach the controversy? The fact that there is no scientific controversy, as explained in Chapter 1, is highly relevant to potential legal outcomes. If the controversy is philosophical, theological, or political, a school could potentially teach the controversy in philosophy, comparative religion, or politics classes, so long as the controversy is taught objectively.[188]

If there were a real *scientific* controversy to teach about, the teach the controversy argument might carry some legal weight. After all, if a school board decided to teach the evidence for and against the existence of non-slushy liquid water on Europa in an astronomy class focused on the solar system, the board would be teaching scientific evidence on both sides of a valid scientific debate.[189] If, however, a board were to require teaching about the role magic might have played in galaxy formation in an astronomy class to teach the controversy between the natural and supernatural, there would be no legitimate scientific controversy. The requirement to teach this controversy would function a lot like the balanced treatment act struck down in *Edwards*.

In fact, teach the controversy arguments are not dissimilar to the balanced treatment approach. To the extent the teach the controversy approach requires the teaching of ID to demonstrate that there is a controversy, the approach runs afoul of *Edwards*, unless ID is science and not religious. Sometimes ID advocates argue that one can teach the controversy simply by teaching that evolution is just a theory and that there are gaps and weaknesses in that theory without teaching ID itself.[190] Unless there is some scientific validity to these arguments, however, these arguments do not support that there is a scientific controversy to teach about, and in fact any controversy only arises when ID advocates and others manufacture it. To the extent that ID is a religious idea, and teaching the controversy is not about teaching a real scientific controversy, there are significant endorsement problems. Moreover, such a policy would run afoul of the holding in *Epperson* that a school board cannot preclude – in this case attack – evolution simply because it is inconsistent with the religious views of certain religious groups. Still, it is helpful to look at both forms of the teach the controversy argument under the *Lemon* test and the endorsement test in greater detail.

The first form of the teach the controversy argument suggests a controversy between evolution and ID or other supernaturally affected views.[191] This is the easiest of the teach the controversy arguments to address under the establishment clause. It requires that ID or other supernaturally grounded ideas be taught directly or rhetorically pitted against evolution. It also assumes that there is a scientific controversy. A social, religious, philosophical, or political controversy would be irrelevant to the scientific merit of evolution. Only if there is a genuine scientific controversy does this argument have any hope of legal success.

As explained in Chapter 1, there is no scientific controversy regarding evolution and ID. Mainstream science does not take ID seriously, nor does it deny the validity of evolution given the vast amount of data supporting evolution.[192] So the first problem with this argument is that there is no scientific controversy to teach about, which leaves only a controversy over the supernatural versus science. Given that courts have treated supernaturally guided theories as religious in this and similar contexts,[193] teaching the controversy would at the very least have the effect of promoting and endorsing religion.

Kitzmiller provides an ideal example of this because in that case the policy and disclaimer suggested the existence of a controversy and offered ID as an alternative to the naturalistic approach of evolution.[194] Of course, the policy and disclaimer were struck down in that case because they endorsed religion and violated the *Lemon* test.[195] Moreover, teaching ID in a science class might violate the indirect coercion test because the school would be promoting a religious subject in an environment where students might feel no real option to dissent, and in fact might be subject to discrimination and retaliation if they do dissent.[196]

We have already looked at the only way in which ID can escape this problem, namely, by redefining science to include supernatural causation. As was explained in Chapter 2, there are significant philosophical and pragmatic problems with such a redefinition. Moreover, the redefinition argument does not provide a workable legal escape from the conundrum. This will be explained in Section V of this chapter.

The second sort of teach the controversy argument poses more interesting legal questions, but the result is the same. Because this argument does not require that ID be taught, but rather that the supposed weaknesses in evolution through natural selection be explained and the existence of alternative ideas set forth, the establishment clause problem is a bit tougher. The argument fails, however, if there is no scientific controversy over the validity of evolution or no scientific recognition of extant alternatives supported by scientific evidence.

What is the purpose of teaching the controversy? Ostensibly, it is to educate students about scientific disagreement. This would all be well and good if there were scientific disagreement regarding the validity of evolution as a scientific theory. Here, however, as with the musical taste example in Chapter 1, the controversy only exists for the ID movement and its allies. From the perspective of every major scientific body to address the issue, mainstream scientists generally, and virtually all state curriculum committees, there is *no* scientific controversy![197] So under the purpose prong of *Lemon* and endorsement, it is essential to consider what the purpose might be in teaching that a scientific theory supported by massive amounts of evidence and accepted by all the major scientific bodies should be the only one in the science curriculum subject to criticism – criticism that is not based in accepted science and is primarily rhetorical (such as the gaps argument).

The purpose would seem to be to undermine evolution, but why? ID advocates are straightforward about this. It is because evolution promotes scientific materialism and methodological naturalism.[198] It excludes the supernatural as a possibility within the scientific context. But would not that be true of any scientific theory? Why single out evolution?

The answer can be found in *Epperson v. Arkansas* – it is because evolution is inconsistent with the religious beliefs of certain religious groups.[199] The fact that ID advocates do not generally acknowledge religious motivations, call Big D God, or promote ID as a religious idea is irrelevant. Even taking all this at face value – and as Chapters 1 and 2 demonstrate there is no reason to do so – ID is still based in the supernatural.[200] Whether the designer is God or the Flying Spaghetti Monster, he, she, or it is supernatural.

The courts have been quite clear that when one promotes supernatural causation as an alternative to science in science classes, one is promoting religion for establishment clause purposes.[201] There is no need to connect it to a specific faith or deity for it to be religion. Therefore the purpose of teaching the controversy is itself religiously grounded and therefore endorses and promotes religion. This argument is even easier to support when one looks at the history of the ID movement and the teach the controversy approach.

Still it is possible that a school board unaware of all this, perhaps one that buys into the ID movement's marketing of creation as nonsectarian science, could act with a purpose to promote critical thinking if it required this form of teaching the controversy. Significantly, this would not save such a policy if its effects endorsed or promoted religion. What would the effect be of using critiques that have been rejected by the mainstream scientific community of a scientific approach that just happens to be inconsistent with certain religious tenets? Consider also that evolution is singled out for such critique among all the scientific subjects studied by students at school.

The result would be to create political insiders and outsiders between those who support mainstream science (even if they also believe in God or supernatural causation generally) and those who see an inherent conflict between mainstream science and religion.[202] The latter is not a necessary dichotomy, as explained in the next chapter, but for present purposes it is enough to point out the effect of the hypothetical policy. The teach the controversy policy would favor those who see a conflict between the natural and supernatural, and an objective observer familiar with the background of this issue would most likely perceive endorsement of religion. Moreover, the primary effect of teaching such a policy would be to promote religion and supernatural causation generally because there is no scientific basis for such a policy. Thus even this latter form of teach the controversy violates the *Lemon* and endorsement tests.

IV. Let the Teachers Teach?

Another potential legal issue arises when a public school teacher wants to teach ID in science classes but the school forbids the teacher from doing so. This question is far easier to address than any of the others in this chapter. This is because public schools have wide discretion in what they can require to be taught and certainly can insist that the state curriculum be taught or that educationally questionable theories, such as alchemy or astrology, for example, not be taught.[203] Academic freedom is the most likely argument a teacher would raise in these circumstances. After all, if the teacher raises a religious discrimination or free exercise of religion argument, the school board wins because the teacher himself or herself would be admitting that the reasons for teaching ID are religious, and the school would have an obvious interest in preventing the resulting establishment clause violation.

The academic freedom defense is more complex at the college and university level,[204] but in the primary and secondary school context, the reality is that teachers do not have the academic freedom to teach whatever they want in whatever course they want.[205] State and local school boards and officials mandate all sorts of course content requirements.[206] Moreover, the establishment clause problem is even greater when a teacher in a primary or secondary school with a captive audience of public school students teaches a religiously grounded approach in a science class – a religiously grounded approach that is rejected by the mainstream scientific community. To the extent that there is any discrimination going on, it is no different than any other circumstance in which teachers are told that some topic is outside the parameters of a class.

If, however, school officials were to disparage the religious commitments of the teacher who wishes to teach ID, there may be a valid issue of religious discrimination under applicable state and federal employment law, depending on how severe and pervasive the actions of the school officials are.[207] The remedy would not be to allow the teacher to teach ID, however, but rather to enjoin school officials from discriminating based on religion and possibly an award of monetary damages.[208] The school district's prohibition on teaching ID in a science class would not be the basis for any such claim, and a court could not require that ID be taught in science classes without violating the establishment clause.

The claim and the remedy would be based on the acts of religious discrimination, and precluding a teacher from teaching ID in a science class is not by itself a form of religious discrimination.[209]

V. Seeding the Academic Forum?

As we have seen, ID advocates have argued that reliance on the current scientific paradigm excludes religious or other paradigms from competing.[210] If such alternative paradigms are to gain any acceptance, the argument goes, it is essential that they be explored by researchers and taught in science classes. This raises the related question of whether alternative scientific paradigms can or should be credited by public schools. Ultimately, any argument that they should be credited draws implicitly, if not explicitly, on the legal concepts of equal access and public forums.[211]

Arguing for inclusion in the realm of science generally requires that one use the tools of modern science. These tools are generally understood as engaging in experiments, or at the very least calculations, that can be repeated and that allow for falsifiability – that is, they could prove that a scientist's hypothesis is wrong.[212] Falsifiability is one key element of modern science, as the famed philosopher of science Karl Popper explained.[213] Still, as we have seen, arguments can be made suggesting that perhaps Popper's definition of science is simply one of many paradigms for science and that there is no superparadigm that allows one to prove the correctness of a given scientific paradigm.[214]

From a legal perspective, however, the possibility of multiple scientific paradigms raises some interesting questions when merged with free speech concerns and the concept of equal access to government forums and programs. ID proponents' arguments that ID should be taught in the public schools – or at least that the "controversy" over evolution should be taught – that they are being excluded from scientific debate, and that they are discriminated against raise the question, from what exactly are they being excluded? As will be explained, there is no issue if ID is not science because excluding it from government-controlled or -sponsored scientific fora would be a natural result of ID not being science. If ID is science, however, it can be argued that excluding it from government-controlled or -sponsored scientific fora raises serious constitutional concerns.

Given that ID proponents make no attempt to falsify their ultimate hypothesis that Big D is involved in creation, do not engage in work that could support ID without presuming Big D's existence, or attempt to falsify their arguments that evolution cannot explain much of what is seen in the natural world, it is obvious that ID is not science under the traditional definition of that term in the era of modern science.[215] If, however, they can argue successfully that the traditional definition of modern science is not the only one, and better yet, that it discriminates against alternative scientific approaches, suddenly ID can claim the mantle of free speech protection even in the scientific context. Of course, this whole argument is question begging because one could make similar arguments about the paradigms for every academic discipline. Thus alchemists can argue for access to the scientific forum, and using a divining rod to find water might have to be included in courses or programs on geology and oceanography. Before jumping to that point, however, it is worth exploring the legal concepts of public forum doctrine and equal access.

A. Public forum doctrine and the equal access concept

Public forum doctrine is relevant when private speakers seek access to government property. Thus, unless private speech is involved, there is no public forum issue. Public forum doctrine arises under the free speech clause of the First Amendment to the U.S. Constitution. It requires that government not discriminate against specific viewpoints or content when it opens its property up to private speakers. There are technically three types of public fora, and they will be discussed in turn: (1) traditional public fora, (2) designated public fora, and (3) limited public fora. ID advocates would need to rely on the second and/or third type of public forum in public school access cases because public schools are not traditional public fora.[216]

A traditional public forum is a space, such as a public park or sidewalk, where a variety of groups or individuals are able to place materials or speak out on a variety of issues.[217] In a traditional public forum, to exclude any message, the government must show a compelling governmental interest for excluding the message and that excluding the message is the least restrictive way to serve the government interest.[218] A designated public forum is one in which the government limits access to the forum to certain categories of individuals such as school students or graduating seniors.[219] While potential speakers may be limited to the category designated, within that group, the same free speech rights attach as in a traditional public forum.[220] A related concept, known as a limited public forum, exists where government limits the topics that may be discussed in the forum,[221] but again, within those topics, all viewpoints must be allowed.[222] The concepts of limited and designated forums are often used interchangeably by courts without any clarification that there is at least a conceptual difference between the two.[223] Government may impose reasonable time, place, and manner restrictions in a public forum subject to a lesser burden,[224] but such restrictions are not an issue in the ID context.

Equal access cases are those in which a religious organization seeks access to government-owned facilities or government-funded fora to which nonreligious entities have access.[225] Equal access cases are in many ways straightforward speech cases. If a school district allows a variety of noncurricular student groups access to school facilities when school is not in session, the school has created a designated or limited public forum.[226] Thus, if a group is excluded from that forum based on the content of its mission or the viewpoint it expresses, the result is content or viewpoint discrimination, depending on the facts.[227] Content discrimination occurs when a government entity discriminates against or excludes an entire subject, while viewpoint discrimination occurs when the government discriminates against speech based on the specific viewpoint involved.[228] For example, it would be content discrimination to exclude all religious speech from a public forum, but it would be viewpoint discrimination to exclude only speech from a Jewish perspective.

Claims of content discrimination in a public forum give rise to strict scrutiny,[229] and thus a government entity charged with content discrimination would need to demonstrate a compelling governmental interest and that its action was narrowly tailored to serve that compelling interest.[230] The Court has suggested that viewpoint discrimination in a public forum is presumed unconstitutional,[231] but it has not clearly addressed this point.[232] There is some support for applying strict scrutiny to viewpoint discrimination, albeit especially strict scrutiny.[233] Regardless, the line between content and viewpoint discrimination is somewhat blurred.

Under current precedent, if religious viewpoints are excluded from a limited, designated, or traditional public forum, even in a sensitive context like an elementary school, the result is viewpoint or content discrimination (depending on whether a specific viewpoint or category of speech is focused on).[234] Treating religion differently in a forum neutrally open to all student groups or subjects is never a compelling government interest because such differential treatment is not required by the establishment clause since the Court has held in access cases that the clause requires religion be treated the same as nonreligion.[235] By assuming that religion must be treated the same as nonreligion, the Court has both set up the claim of viewpoint discrimination and answered the compelling interest defense to that claim.[236]

So what does all this have to do with ID? Assuming that ID is considered religious, a fair assumption given its supernatural focus, history, and basic tenets, one way ID advocates may get around establishment clause concerns is to assert that they are being excluded from a public forum. This argument does not help, however, because a school's curriculum or a specific class is not a public forum.[237] If a student club focusing on ID or even creationism was precluded from meeting on campus before or after school hours, and that school opens its campus to non-curriculum-related student groups, there would be a free speech violation under the public forum doctrine.[238] But of course, it is access to science classes that ID advocates seek, and because those classes are not public fora of any sort, excluding ID does not give rise to a free speech violation. As was noted earlier, schools control what is taught in classes, and the classes themselves are clearly under the direct control of the school and its staff. These are not public fora for private speech.

Another way ID advocates might try to use public forum doctrine is to assert that science itself, or the determination of what counts as science in public schools and government funding fora, is essentially a public forum. Therefore excluding nontraditional scientific approaches is a form of viewpoint discrimination. The most obvious response to this is that science classes in the public schools and government funding for scientific research are not public fora for private speech. Even if they were, they would be designated or limited public fora, and to be part of the designated group that has access to the fora, one must be teaching or engaging in science. Science is the criterion for entry into the forum. The ID response would be that "ID is science," so why exclude it when it is just a different paradigm for science? That question will be answered in the next section.

B. Paradigms, equal access, and public fora

Given existing legal precedent and scientific evidence, the best argument ID advocates have for including ID in the scientific – as opposed to theological and/or philosophical – realm is to combine equal access–public forum arguments with relativist epistemology. The argument would go something like this: ID has a place at the scientific table (regardless of its merit under currently governing scientific paradigms) because it is a plausible paradigm for science and ID advocates have engaged in research that supports it. Given this, excluding ID from the scientific forum is a form of viewpoint discrimination because ID is an alternative explanation of the nature of complex life-forms, and denying it access gives a privileged position to evolutionary theory and scientific materialism.

The problems with this argument are manifest. First, as explained earlier, most of the situations where this argument will be made are not public fora, or even limited or designated public fora. A school's curriculum is not a public forum for all theories no matter how far afield of current knowledge.[239] Moreover, to acknowledge even that the science curriculum may be a limited public forum would do ID advocates little good because it would be limited to "scientific" theories, which of course begs the question.

To respond, ID advocates would have to use relativist arguments about epistemology (how we know things) as a practical tool, arguing that ID is just an alternative scientific paradigm and that there is no way to judge it inferior to the alternatives without engaging in viewpoint discrimination. As noted earlier, the results of this approach would turn public school curricula into a free-for-all for every possible paradigm in every discipline no matter how unacceptable or discredited those paradigms are in the relevant discipline and regardless of whether the leading thinkers, researchers, and associations have rejected the alternative paradigm. A child's school day could consist of homeroom (time for students to hang upside down to gain better flow and balance), biology (where ID is taught), chemistry (where alchemy and ethers are taught), English (where *Mad* magazine is the major text because it is "great social commentary"), history (where Bible history or the depravity and destructiveness of Western societies is taught), art (where the work of Maplethorpe and Andres Serrano are taught, with an extended focus on Serrano's "Piss Christ"), and finally physical education (where the school of hard knocks approach to dodge ball is the main focus). Though this example is a bit extreme, these course possibilities exist if we declare equal access or public forum doctrine applicable to courses or disciplines and then make the arguments necessary to include ID as science.

The free speech arguments prove too much to help ID advocates, and the combination of free speech arguments with scientific relativism leads to bad legal doctrine, acontextualized use of potentially valid metaphysical arguments, and ridiculous practical results. Also, for the reasons stated elsewhere in this book, the arguments that privileging evolution establishes a religion of secular humanism or that the denial of the "right" to teach ID as science in public institutions denies free exercise rights fair no better.[240] In the end, because of ID's failure in the proof game and the inconsistency between the absolutist commitments inherent in ID and the use by ID advocates of relativist arguments, ID can gain no legal traction for inclusion as science even when the best potential arguments for such inclusion are put forward. The public forum issue will be revisited in the college and university context in Chapter 6.

4 Theistic Evolution

I. Introduction

Intelligent design (ID) faces serious opposition not only from science but also from religion. Many people, and many organized religions, accept theistic evolution; the belief that God exists (in the case of monotheists) and that modern science, including evolution through natural selection, is valid.[1] God and science need not conflict.[2] Many religions, from the Roman Catholic Church to Conservative and Reform Judaism, accept theistic evolution,[3] as do many scientists and others throughout the world.[4] Theistic evolution is a serious threat to ID's legal and philosophical arguments as well as its marketing appeal. Yet most people do not know much about the differences and conflicts between ID and theistic evolution. This can lead to confusion, with some people believing ID is theistic evolution and perhaps vice versa. Section IV of this chapter will address the significant differences between these two ideas, the threat theistic evolution poses to ID, and why ID advocates often try to ignore the topic.

First, it is worthwhile to explore a problem that ID advocates and some creationists alike try to manufacture, that is, the view that one cannot believe in the Bible or God without accepting a literal or quasiliteral reading of Genesis, or in the case of ID advocates (who often share this underlying view but do not acknowledge it), the view that one cannot accept supernatural causation (including God) if one accepts modern science's methodological naturalism because the latter precludes the former.[5] The discussion will focus initially on more literalist creationists because arguments regarding literalism, especially relating to the human capacity for interpretation, can as you will see, apply to ID as well as creationism.

II. Why Design Beings Who Can Interpret and Explore If Everything Is Predetermined?

The idea propounded by many creationists that the Bible (usually in one of its weaker English translations) must be read literally, including the Genesis flood, Adam and Eve, and the like, reflects a remarkable hubris. These folks believe they know what God intended by the words contained in Genesis as translated from its original language and cultural context. Of course, the same Bible states that God gave Moses the Torah, which includes the book of Genesis in both Christianity and Judaism, at Mt. Sinai. No one alive at that time would have known Noah or any of the other characters in the early part of Genesis. They may have been familiar with some of the epics and stories that were handed down through various ancient civilizations, such as the epic of Gilgamesh, which tells of a great flood.[6] Moreover, cultures in that area at that time often used stories to explain the unknown or to teach valuable lessons.[7] Some of these were taught as religious truth, but others were intended as historical epics or to explain a moral principle.[8]

How might a supreme being teach such a culture about the importance of using knowledge wisely, the value of faith, and the folly of extreme sibling rivalry? Perhaps the stories of Adam and Eve, Noah, and Cain and Abel would be helpful. Especially since the people of Israel in Moses's time may have already been familiar with the epic of Gilgamesh or some of its later versions.[9] None of this is inconsistent with science. No worldwide flood need have occurred, Adam and Eve need not have literally been the first people, and Cain and Abel may simply be archetypes of the jealous rivalries that have unfortunately affected society and some families throughout the ages.

Science explains the natural world as it is and was. One may fully accept that God created the material world and set natural processes into play, and as Kenneth Miller has explained, in doing so fostered free will because in our material world, there is always an element of chance and free will.[10] Under this view, Genesis might be what some would call a teaching moment. One cannot, however, take Genesis literally – especially in some of the weaker translations – and accept the material world as understood by science because it is clearly far older than a literal reading of Genesis would allow.[11] Moreover, the evidence for evolution is overwhelming, and the evidence that there was no worldwide flood is equally overwhelming (regional floods are a different matter).[12]

If one presumes to know the will and intent of God, as the literalists do, one eliminates much of modern science, but if one accepts that God may not have intended for Genesis to be taken literally, a world of possibilities opens up – possibilities that allow science and Western religions to reconcile. As will be seen later in this chapter, science and some Eastern religions may be even easier to reconcile.

God, being God, could have sensibly used such stories to teach various lessons. Why make such an assumption? Many theologians and philosophers have, for a variety of reasons. The reason I suggest this here is to demonstrate the hubris inherent in biblical literalism. I do not assert that my suggested explanation is correct, only that it is another – one of many – coherent alternatives to literalism. Few would argue against the proposition that humans have the capacity to reason and interpret. We must interpret the world around us, interactions with others, texts, and even art. In doing so, we are affected by our traditions, which can shape the horizons through which we view the world.[13] We project our preconceptions into every interpretive act.[14] The things we interpret also have potential meanings based on what is known about them.[15] When we can converge our horizon with that of what we are interpreting, meaning occurs.[16] In many scenarios, this happens almost automatically, especially when we share the traditions of whatever we are interpreting.[17]

This has serious ramifications for biblical interpretation. First, those who argue for literalism must assume that God is either unaware of our ability to interpret or that though God gave us such a great capacity for interpretation, God expected us to apply it everywhere but to Genesis. Second, it ignores the central point of how we interpret, which is that it is our traditions and contexts that embed the horizons through which we interpret. Though we may share something with the desert tribes who lived more than 3,000 years ago, spoke a different language, had vastly different technology and social ordering, and used allegory and storytelling as central means to impart morals and wisdom, we clearly do not have the same horizon. Why would God be unaware of this?

If God gave a book to be understood throughout the ages, and God gave us the ability to interpret, why would it be expected that we would take it literally? In fact, requiring such literalism would mean the Bible could be more changeable than a nonliteralist approach would require because every era and culture would be affected by different traditions that can cause variations in the interpretation of the Bible.[18] We see this even in the creationist world, with some creationists accepting an old Earth, others only a young Earth, others a middle-aged Earth, and a variety of views on the viability of evolution ranging from the idea that evolution is completely unreal to the idea that it is only unreal as to humans.[19]

The description of interpretation sketched out previously is not the only one. It comes from a philosophy known as *philosophical hermeneutics* and specifically from a philosopher named Han-Georg Gadamer.[20] I realize that it is not without philosophical controversy. That broader debate is beyond the scope of this book. There are numerous views on how humans interpret, yet virtually no philosopher, linguist, or social scientist would suggest that we do not interpret – interpretation is a given part of human existence. So why give us this capacity if we do not use it, and why should a God who is aware of the abilities he gave us not expect us to use them, and why would he not set forth his message in such a way that it would work throughout the ages as human knowledge evolved?

Literalism does not allow for this because it does not generally acknowledge that literalism itself is a form of interpretation that may require round belief pegs to fit into square human knowledge holes. Theistic evolution shows that one can have faith and accept increased human knowledge without conflict. Theistic evolution allows for a square peg of faith to fit into the square hole of knowledge about the material world.

This makes theistic evolution a significant threat to literalist creationism and also to ID. After all, not only does theistic evolution suggest that science and faith are consistent, which implicitly rejects literalist views of genesis, but it also suggests that scientific materialism itself is perhaps God's methodology. Therefore, theistic evolution conflicts directly with a central tenet of ID, namely that scientific materialism is inconsistent with supernatural causation.

Not all theistic evolutionists advocate that scientific materialism is God's methodology. Some theistic evolutionists argue that religion and science are simply separate modes of inquiry and therefore one can accept scientific evidence without losing faith. Yet this latter argument too acknowledges that evolution is real, and thus literalist readings of Genesis must fall to the scientific evidence. This does not require one to abandon faith, but it does require one to abandon literalism, which itself is ridden with assumptions, translation errors, and presumptions about God's intent.

In the end, however, the simple reality is that theistic evolution demonstrates that it is possible to reconcile faith with methodological naturalism and scientific materialism. In other words, one can have strong faith, accept scientific evidence, and reject the core underpinnings of creationism and ID – not exactly an idea that creationists and ID advocates want embedded in the public conscience.

III. Marketing Dualism or Minimizing God's Possibilities

If one assumes that faith and evolution can be consistent, ID becomes irrelevant. There is no need to bypass evolution or presume a Big D of the gaps. For ID to have any real traction, there must be tension between evolution and supernatural causation. If evolution is the method a given supernatural entity chose, than the method, but not the entity, can be tested scientifically. ID presumes a tension between supernatural forces and scientific materialism.

And this is key: it presumes that the tension is a scientific tension! If the tension is metaphysical rather than scientific, so that some people see naturalism as an end in itself, while others accept the results of methodological naturalism such as evolution but believe that this does not preclude a supreme actor, ID's raison d'être disappears. Simply put, ID needs dualism.

ID must pit faith (the supernatural) as demonstrable through science against naturalistic science to market itself as an alternative to materialism or as a solution to the assumed tension between science and faith. Thus ID advocates have a vested interest in supporting the notion that science excludes the supernatural from the world of possibility. Of course, one might argue, as some theistic evolutionists do, that science does not preclude the supernatural, but by its nature the supernatural cannot be proven by science. Science and faith are separate modes of inquiry, but when in conflict, one must choose the evidence of science or at the very least acknowledge that rejection of that evidence is based on faith rather than alternative scientific proofs.

By marketing this dualism, however, ID not only assumes that the supernatural is scientifically demonstrable – somewhat ironic given that it is called supernatural because it is beyond the natural realm, and the natural realm is the focus of modern science – but it also relegates the supernatural to the realm of gap filler, intellectual caulking, if you will. Yet one can fully accept the natural world that science explains and still believe that there is a place for faith and the supernatural. One simply must accept that the latter beliefs are not based in science or the natural world and therefore are not likely to be persuasive to those who seek natural evidence to explain the world. Unless one believes it important to market one's faith and belief in the supernatural, that is, unless one seeks to proselytize, there would be no need to support faith through science, a mechanism poorly tailored to do so.

Ah, but that is the rub: ID is all about proselytizing. The Wedge Document makes this clear,[21] as do numerous other pieces of evidence.[22] If everyone were to accept that faith in the supernatural cannot be demonstrated through science, materialism would reign. ID advocates[23] – and some of their staunchest opponents in the atheist world[24] – believe this would be the end for religion and especially for ID advocates' favored brand of Christianity.[25] Thus, to market creation in court and in society, ID advocates believe they must propose a scientific alternative to scientific materialism and methodological naturalism.[26] This of course assumes that faith needs such proof to be persuasive, and in this it may be fair to accuse ID advocates of underestimating the power of faith and God in their zeal to market creation. This is where theistic evolution and plain old creationism demonstrate a huge contrast to ID and creation science.

Traditional creationists assume that if God and science conflict, God's will and God's word in scripture win out. Theistic evolutionists assume that the conflict between science and faith itself is an artificial dualism and that God and evolution can coexist. Both these groups appear to believe that God's appeal can be argued on its merits without claiming that God is a scientific, as opposed to a supernatural, fact. ID advocates may argue that this is all well and good, but it overlooks the fact that science itself proselytizes and therefore faith must be able to challenge materialism on its own terms or faith may lose a major forum for persuasion – modern society itself (and all that modern society entails).[27]

In a way, this is true. Every idea proselytizes in some manner. Some people may or may not be persuaded by a variety of ideas on a variety of subjects. Science, however, to the extent that it persuades people, does so on the basis of evidence gleaned from experimentation.[28] Whatever assumptions underlie science generally, it would not have its current persuasive force if it was not so darned supported by evidence!

Look at the world around you, and evidence of science's success can be seen everywhere, from the combustion engine to medicine to the molecular structure of the roads we drive on to the data processing that enabled me to write and Cambridge University Press to publish this book. Evolution, too, is supported by overwhelming evidence, as explained elsewhere in this book.

Of course, a person may look at the world around him or her and see God's hand as well. There is nothing wrong with this. The difference is that while science can demonstrate how a tree functions and even from what sorts of earlier trees it evolved, science cannot explain why God would make a tree or what role God had in tree evolution (or spontaneous creation, if one so believes). Science is supported by physical and/or mathematical proof and in fact at its very core acknowledges that answers may change as more evidence arises.

Faith may be supported by reasoned proof, often deductive or existential, but in the end, faith is a different mode of inquiry. Yet when ID advocates seek to use science to prove faith, they enter science's proof game, and quite simply, they are not up to the task. Their faith assumptions affect every argument they make; their need to proselytize affects their faith assumptions; and in the end, ID is just a sophisticated form of religious apologetics.

When ID advocates argue that science proselytizes, they are not wrong – they simply miss the point. Every idea on every subject proselytizes in a way, but to the extent that science persuades/proselytizes, it does so not because its goal is proselytizing but because its goal is explaining the natural world through physical and/or mathematical evidence. It is persuasive because it is supported by a lot of physical evidence, to borrow a legal term (for those unfamiliar with the term, physical evidence is exactly what it sounds like: a gun, fingerprints, DNA, a document). ID, however, has not demonstrated any such credible physical evidence. It must rely on purely circumstantial evidence, circumstantial evidence that is undermined by the physical evidence of science and potentially by the alternative circumstantial evidence posed by theistic evolution and traditional creationism. Of course, that is the difference between science and ID: science's goal is to explain the natural world, whereas ID's goal is to persuade people of Big D.

IV. Theistic Evolution versus Intelligent Design

Theistic evolution openly acknowledges belief in God and in science. This might seem surreal when compared to ID, which rarely acknowledges that Big D is God – even if that conclusion seems rather obvious – and which questions well-supported scientific evidence because of its inconsistency with Big D, whatever he or she is. Of course, theistic evolutionists do not seek to have theistic evolution taught in the public schools, nor do they seek to gain scientific acceptance for the theistic aspects of the approach, and most important, they do not claim science as a means to prove their religious beliefs or the supernatural more generally. Theistic evolutionists accept science, and many believe that their faith can coexist quite well with science even as science itself evolves.[29] Therefore, because they lack an agenda to count the theistic aspects of theistic evolution as science, they need not repackage what they believe to market it so that it will be legally acceptable in science classrooms or viewed as a "scientific" alternative to mainstream science.

By contrast, ID seeks to claim the mantle of science for its supernatural claims; to include its tenets or, at the very least, its arguments against evolution in public school science classrooms; and to gain public acceptance as science. To survive, ID must win the marketing battle at some level because, as we have seen, it is unlikely ever to win the scientific battle unless it changes the ground rules for science. Thus, if ID can rally those who already share its horizon that faith and evolution are inconsistent and use fancy terminology and pseudoscience to persuade others, it can survive numerous scientific defeats. After all, marketing strategies can work for products that have significant flaws or make promises that the product cannot live up to.

That is, unless it becomes widely known that there is an alternative that allows for faith and science to be reconciled and which suggests that a God who used evolution as a tool for creation acted far more efficiently, and less cruelly, than the one that biblical literalism, or even ID, demands. After all, if God directly designed and created all complex life, it is awfully cruel to have the vast majority of species that have ever lived go extinct.[30] Moreover, if it were to become widely understood that our capacity to interpret – if one believes God had a role, or *the* role, in creating us – works against literalism, and against ID, it would become more plausible simply to accept science as evidence of the methods God used, methods that include evolution, given the overwhelming scientific proof supporting it.

In a way, this causes theistic evolution and traditional creationism to be strange bedfellows on one major point. Neither are satisfied with minimizing God into some sort of gap filler or unacknowledged designer. Theistic evolution and creationism may disagree at their very core on evolution, but they agree that God is relevant because God is beyond the realm of the material world, not just some marketing ploy like Big D. If theistic evolution were to catch on – and it is catching on – ID would lose a significant potential market: those who seek to reconcile their faith with science. This puts a squeeze on ID, which already faces an uphill battle in convincing some true creationists of ID's watered-down creationist assertions.[31] Those most likely to share a horizon with ID would be those predisposed to believe in God and seek some reconciliation between that belief and science. Since creation science can fill this role for many literalists, theistic evolution, which can fill this role for other people of faith, especially those who understand a bit about science, is a significant threat to ID's long-term marketing goals. This problem is exacerbated by the fact that so many, and such a diverse group, of organized religions and people of faith have spoken in favor of theistic evolution,[32] as have major scientists and theologians.[33]

ID is mostly supported by highly evangelical religious entities that appear to see it for the apologist strategy it is and by a number of public or political figures.[34] It is not supported by any mainstream scientists who have significant credibility in the scientific community.

How has the ID movement responded to this threat? Frequently by ignoring it. It is rare for ID advocates to acknowledge theistic evolution. When they do, it is usually in a negative and oversimplified manner.[35] ID does, however, benefit to a certain extent from confusion about theistic evolution; that is, many people confuse the two concepts, and this may help ID advocates in public opinion polls and gaining public support. Not discussing or debating theistic evolution, while focusing on those mainstream scientists who venture beyond science to address metaphysical claims, is a wise strategy for ID advocates. By focusing on the artificial dualism between faith and science that ID advocates and creationists manufacture, ID advocates can marginalize or ignore the very approach that demonstrates this dualism is a false dualism or at the very least an unnecessary dualistic assumption.

Theistic evolution exists both in what one might call strong and weak forms (although which is which may depend on one's perspective). The strong form of theistic evolution accepts that God and evolution are consistent; that science generally answers only questions about the natural world, not broader questions of meaning that have long vexed humanity; and for some "strong" theistic evolutionists, evolution demonstrates the sophistication and greatness of the creator. Of course, theistic evolutionists are generally open and clear that questions of meaning and God are answered by reflection and/or faith, not by science. This concept is well captured by Kenneth Miller in his seminal book *Finding Darwin's God: A Scientist's Search for Common Ground between God and Evolution*:[36]

> A nonbeliever, of course, puts his trust in science and finds no value in faith. And I certainly agree that science allows believer and nonbeliever alike to investigate the natural world through a common lens of observation, experiment, and theory. The ability of science to transcend cultural, political, and even religious differences is part of its genius, part of its value as a way of knowing. What science cannot do is to assign either meaning or purpose to the world it explores. This leads some to conclude that the world as seen by science is devoid of meaning and absent of purpose. It is not. What it does mean is that our human tendencies to assign meaning and value must transcend science, and ultimately must come from outside of it. The science that results, I would suggest, is enriched and informed from its contact with the values and principles of faith. The God of Abraham does not tell us which proteins control the cell cycle. But He does give us a reason to care, a reason to cherish that understanding, and above all a reason to prefer the light of knowledge to the darkness of ignorance.[37]

Miller's eloquent discussion is echoed in statements by numerous religious bodies, including the Roman Catholic

Church, a number of Protestant churches, many in the Jewish community, the Dalai Lama, and ecumenical organizations such as the National Council of Churches.[38] A recent statement on *Science, Religion, and Teaching of Evolution in Public School Science Classes*, released by the National Council of Churches Committee on Public Education and Literacy, noted the following:

> ***What is science?***
>
> Science is the study of the material, processes, and forces in the natural world. Science is not about belief; it is about how things work. One cannot "believe" in science or "believe" in evolution. Science is about the exploration of natural causes to explain natural phenomena. Science is empirical, which means that questions of truth are established through experimenting and testing. There are no absolutes in science; all issues are open to retesting and reconsideration.
>
> ***What is religion?***
>
> Religion is about belief, meaning, and purpose. According to the Episcopal Church, the stories of creation in Genesis "should not be understood as historical and scientific accounts of origins but as proclamations of basic theological truths about creation. 'Creation' in Holy Scriptures refers to and describes the relationship between God and all God's wonderful works." Religious truths are evaluated by an appeal to authority, by contextulization in history, by their philosophical coherence, even by their psychological and emotional resonance with life and experience.[39]

Rabbi Jeremy Kalmanofsky, from the United Synagogue of Conservative Judaism, eloquently demonstrates the many significant problems with the artificial dualism created by ID advocates, creation scientists, and their most vociferous atheist opponents such as Christopher Hitchens and Richard Dawkins:

> Truth is, many religious myths must become metaphorized, as science reveals that they describe something other than factual reality. . . . But even sacred metaphors remain valuable, because they point toward spiritual truths, which, while they cannot flout science, are not limited to what is provable in a lab.
>
> For instance, everyone knows love is real, though not in the way a quartz crystal is real or even in the way an electron is real. Similarly, we may speak of God creating people in the divine image, because this language expresses the remarkable – perhaps unique – human capacity to understand the universe's objective physical order and implicit moral order.[40]

Similarly, the United Methodist Church has issued an official statement on the subject, called *Science and Technology*:

> We recognize science as a legitimate interpretation of God's natural world. We affirm the validity of the claims of science in describing the natural world and in determining what is scientific. We preclude science from making authoritative claims about theological issues and theology from making

> authoritative claims about scientific issues. We find that science's descriptions of cosmological, geological, and biological evolution are not in conflict with theology. We recognize medical, technical, and scientific technologies as legitimate uses of God's natural world when such use enhances human life and enables all of God's children to develop their God-given creative potential without violating our ethical convictions about the relationship of humanity to the natural world. We find that as science expands human understanding of the natural world, our understanding of the mysteries of God's creation and word are enhanced.[41]

Perhaps the most famous statement regarding the compatibility of evolution and faith came about when Pope John Paul II gave his *Message to the Pontifical Academy of Sciences: On Evolution* on October 22, 1996. He said in part,

> Before offering a few more specific reflections on the theme of the origin of life and evolution, I would remind you that the magisterium of the Church has already made some pronouncements on these matters, within her own proper sphere of competence. I will cite two such interventions here.
>
> In his encyclical *Humani Generis* (1950), my predecessor Pius XII has already affirmed that there is no conflict between evolution and the doctrine of the faith regarding man and his vocation, provided that we do not lose sight of certain fixed points.
>
> For my part, when I received the participants in the plenary assembly of your Academy on October 31, 1992,

> I used the occasion – and the example of Galileo – to draw attention to the necessity of using a rigorous hermeneutical approach in seeking a concrete interpretation of the inspired texts. It is important to set proper limits to the understanding of Scripture, excluding any unseasonable interpretations which would make it mean something which it is not intended to mean. In order to mark out the limits of their own proper fields, theologians and those working on the exegesis of the Scripture need to be well informed regarding the results of the latest scientific research.[42]

These are just some examples of the many similar statements made by religious movements, leaders, individuals, and theologians. The statement by Kenneth Miller demonstrates that it is also a view held by many scientists because he is certainly not alone among mainstream scientists in holding the view that religion and evolution are compatible. Theistic evolution is a powerful concept that appeals to our rational understanding of the natural world and the role of science in explaining it and to our quest for meaning in that world.[43] As the next chapter demonstrates, some suggest that the latter quest is either unnecessary or subsumed in the first, but it is precisely the rejection of the all-or-nothing, science-or-faith approach that makes theistic evolution so appealing to many people of faith who cannot ignore the obvious importance and transformative power of scientific knowledge.

Yet what about those who accept scientific knowledge and may not be strongly affiliated with a given faith or even a given view of God, that is, people who may believe there is something greater out there (some may believe this is God) but do not give it much thought? This is where "weak" theistic evolution comes into play. This sort of theistic evolution is common among those who are spiritual to some degree but not religiously affiliated.

I refer to it as weak theistic evolution because it need not focus on ultimate meaning or God's will or interaction with nature. It is simply the idea that the spiritual, including notions of a deity or deities, and science are reconcilable. Perhaps evolution is the method God used to create, or perhaps such spiritual understanding is beyond our comprehension or even our need to comprehend, but evolution is a fact of nature regardless. This form of theistic evolution may be reflected in the ideas of those who have deeply thought about the interaction of science and the spiritual, or it may simply be the intuition of educated people who accept science but are unprepared to abandon faith. Either way, like its strong counterpart, weak theistic evolution rejects the false dualism fostered by ID advocates.

Simply put, theistic evolution in either form is an extreme danger to ID. Interestingly, it also poses a challenge to those, such as Richard Dawkins, who assert that belief in evolution leads implicitly (or explicitly) to atheism.[44] This latter view is the subject of the next chapter. Theistic evolution raises a challenge to these two extremes precisely because it refuses to take a rigid approach. It is not biblically literalist, concerned about proving God or the supernatural through science as ID advocates and other religious apologists are, or focused on the idea that science is the only way to understand existence.

Existence is an awfully big mountain, and science beautifully explains much about the structure, makeup, flora, fauna, and other physical properties of the mountain, but it does not explain why (as opposed to how) the mountain is there or whether the actions of the flora and fauna have some deeper meaning. Perhaps one day, it will be able to do so in purely naturalistic terms, and if so, as explained in the next chapter, fascinating new questions will be raised for people of faith, but theistic evolutionists would be best equipped to *honestly* confront such a situation. For now, however, theistic evolutionists demonstrate that there are many horizons and vantages from which to view the mountain of existence and that each view may enrich the others so that the scientific vantage enriches religion and the religious vantage helps answer metaphysical questions that science cannot currently answer.

To ID advocates who believe that the scientific naturalist view of the mountain must block other views of the mountain, dualism is a tool of great power. For if they can convince people that one must choose Darwinian evolution or the supernatural, they may create enough cognitive discomfort to cause people who see the value in the scientific view of the mountain to reject the scientific view for fear that accepting it would require a rejection of God and the supernatural. Dualism, particularly in its basic deductive form, is a powerful weapon in the hand of a religious apologist.

Theistic evolution, if seen as an option by those ID advocates put to the choice between faith and evolution through natural selection, may be an option many people choose. For ID advocates, this is not acceptable, and therefore ignoring theistic evolution, treating it as a copout, or even claiming commonality with it are potentially effective ways for the ID movement to keep its dualistic train on the track toward public acceptance. It is not surprising, then, that ID advocates pay little public attention to theistic evolution.

5

Dawkins's Dilemma

I. Richard Dawkins on Evolution and Religion

We have already addressed intelligent design's (ID's) inability to gain support in the scientific community and the fact that ID can gain no practical benefit from scientific relativism. Moreover, we have considered religious approaches that allow one to accept the overwhelming scientific evidence for evolution through natural selection without rejecting belief in God. The apparent clash between science and religion is necessary to the survival of ID and creation science because if there is no inherent inconsistency between science and religion, there is no need for religion to play the proof game – that is, there is no need for a Big D of the gaps. ID advocates must manufacture the clash between religion and science to market design as *the* alternative to scientific materialism. Ironically, as Kenneth Miller has explained, they have been aided in this task by some of the brightest minds in science.[1]

One of the most scathing critics of ID is leading evolutionary biologist and scientific advocate Richard Dawkins, Charles Simonyi Professor of the Public Understanding of Science at Oxford University. Dawkins's attack on the presumption-laden – and scientifically barren (from the perspective of normal science) – nature of ID is supported by much of the analysis in this book. Yet Dawkins goes beyond scientific analysis of ID and enters the debate over religion and specifically the existence of God or gods more generally.[2] His entry into this debate should be welcomed by all thinking people as his work asks excellent questions,[3] but as will be seen, his answers sometimes rely on obvious precommitments.

Dawkins asserts quite simply that God is a myth, or at least exceedingly likely to be, and that religion is more damaging than it is helpful to the world and those who occupy it.[4] Many have responded to his arguments, which some have characterized as a form of evangelical (in the proselytizing sense) atheism or atheist apologetics.[5] The latter characterization essentially turns his argument on its head. His point is that answers to ultimate questions are just as amenable to scientific proof as other questions, even if science may not yet be at a level that can definitively answer those questions (science cannot yet answer the question of how humans can travel interstellar distances in a single lifetime, but at some point it may).[6]

Dawkins acknowledges that science has not definitively answered the question of whether God exists but argues that such absolute proof is unnecessary to convince a rational mind because one can still consider the probability that God exists using current scientific knowledge.[7] All this makes a great deal of sense, but then he begins what amounts to a detailed argument against the existence of God that is often more deductive or anecdotal than scientific.[8] This is where he falls into something of the trap into which ID advocates have intentionally walked. Dawkins begins with the assumption that God is very improbable and bases this presupposition heavily on a variety of arguments.[9] Several of these arguments are in a sense self-fulfilling in that they rely on how one frames the problem rather than a purely scientific approach.[10] In this regard, we have already seen Kenneth Miller's very different framing of such questions.[11]

Perhaps Dawkins's most persuasive argument is a scientific argument, and a darned good one, too. I will frame it briefly as follows: given that complex organisms evolve, how is it possible that anything complex enough to design or put the evolutionary process into play could exist before all else?[12] Those who advocate design might at first glance believe this argument helps them, but it does not. After all, it necessarily raises the question: Who designed the designer?

In other words, if all complex life-forms are demonstrably designed due to their complexity, then anything complex enough to design would have to be designed![13] Therefore, as a demonstration of the patent problems with ID,

Dawkins's argument is one of the best yet. But does it prove that the probability of God in whatever form is any more exceedingly low than the probability of God's nonexistence? This is the main purpose to which Dawkins puts this argument.

Dawkins seems to begin with the presumption that the probability of God's nonexistence is greater than the probability of God's existence, and he sets off on his analysis from there.[14] He acknowledges numerous times that many scientific phenomena, including evolution, are exceedingly unlikely, but of course, as Dawkins eloquently demonstrates, evolution is real.[15] Given our current level of technology and the template for evolutionary biology laid out by Charles Darwin,[16] science has demonstrated biological evolution through the fossil record and through genetics.[17] The ability to date fossils as precisely as we do today or to analyze evolutionary change through DNA and RNA would have seemed exceedingly unlikely before Darwin.[18] Yet today we take these scientific advances for granted.

The key is that science can be used to explore the natural world and may disprove earlier presumptions. Yet Dawkins does not apply the scientific method to his presumption about God.[19] If he did, the best he could say is that if there is a God, God must have evolved somehow or must be the result of processes that we do not yet understand (and these could be natural processes that would simply seem supernatural to us given our level of technology). Perhaps this seems to Dawkins, given his preconceptions, to increase the probabilities against God, but given our current level of knowledge, this remains a presumption.[20] Similarly, theistic evolutionists like Kenneth Miller and I must rely on presumptions (regarding "theistic" not "evolution") until more data are in.[21] Given our preconceptions, we look at the same evidence and see that science and God can coexist quite consistently and that the probability could tilt just as easily toward, as against, God.[22]

Of course, both these conclusions, Dawkins's and mine, are affected by our preconceptions on ultimate questions.[23] These preconceptions can limit or expand the horizons of what we can envision.[24] But what if we put those preconceptions to the side and look only at the scientific evidence (which I suspect Dawkins, Miller, and I would all support with equal verve)? The best we could say is that the jury is out. Science does not prove that God exists, and it does not prove that God is a myth.[25] At some point, perhaps, science will demonstrate a real skew in probabilities suggesting whether God exists. As Dawkins points out, one does not need absolute proof one way or the other for science to inform us on these issues.[26]

My differences with Dawkins are not over the value of probabilities or the ridiculousness of seeking absolute proof as the only means of demonstrating the status of God but rather on what current science can teach us. In this sense my position is theistic evolution's mirror image of Dawkins's position. He suggests that if there ever is scientific proof demonstrating a strong probability that God exists, he would accept it.[27] I agree. If there is ever scientific proof demonstrating a strong improbability of God, I would accept that proof. My point is that in the short run, any assessment of such probability is speculative given the effect our preconceptions have on the limited data that exist. In fact, one of the most interesting implications of Dawkins's work – although one that was almost certainly not intended – is the possibility that God is not supernatural at all but rather a natural phenomenon (if in existence at all).[28]

One of the great assumptions made by all involved in the debate over God or gods is that whoever or whatever God is must be supernatural. After all, this is why the ID and creationist movements have focused so heavily against what they call scientific materialism and methodological naturalism and why others, such as Richard Dawkins and Christopher Hitchens, assume that religious perspectives are inherently bunk relied on by people throughout the ages to fill psychological needs[29] or simply a negative by-product of positive naturally selected traits.[30] What if – and this is purely hypothetical – what we call God is just so advanced that we see it as supernatural to the extent that we see it at all? Dawkins points out this phenomenon. If something is so advanced that less advanced cultures see its acts as supernatural, that is more the result of the less advanced culture's perspective and scientific limitations than reality.[31]

To counter Dawkins's probability argument (only for the purpose of demonstrating that there may be more possibilities out there than Dawkins or ID advocates suggest), consider that string theory has posited that there are numerous universes (as has quantum theory, but the rub is literally why I focus on string theory here).[32] These universes may rub against each other in a variety of ways, and these interactions can destroy universes or spawn more universes.[33] If this is so, there is no reason to assume that all universes must adhere to the same physical laws,[34] but let us assume for a moment that they do. Therefore complex lifeforms, even those based primarily on energy, must evolve.[35] A species or entity may have evolved in a different universe and broken through to ours or even caused the reaction that gave birth to ours. Although it is far beyond current science's ability to evaluate, this would be a natural phenomenon. Of course, it leaves the question who started the universe in which that entity evolved, but that just places us back at square one with our preconceptions and the boundaries of what may be possible through natural means. String theory, and its relevance to the ID debate, will be addressed in more detail in Chapter 6.

The point is that once we drop our preconceptions regarding the existence or nonexistence of God, and God's supernatural or natural origin, other possibilities exist. The problem with the probability argument, once divorced from preconceptions, is not that Dawkins is wrong. In fact, he is almost certainly correct that science can determine over time the probability of God. Rather, it is that he subsumes what science can currently demonstrate within his own preconceptions. In this sense he falls into the trap the designers happily run into (often without realizing that it traps them in the proof game). The major difference between Dawkins and ID advocates is that Dawkins is exceedingly honest about his views (openly demonstrating his preconceptions),[36] and of course, Dawkins is not a religious apologist; he is a scientist.

II. Dawkins and the Proof Game

Unlike the ID and creation science movements, many theologians and theistic evolutionists have argued that science and religion are simply different modes of inquiry.[37] Thus science cannot answer the ultimate questions that religion addresses, and religion is unequipped to prove its tenets through science.[38] Until reading Dawkins's *The God Delusion*,[39] I generally adhered to this view, while understanding as a practical matter that science could certainly disprove the literal truth of some religious positions, for example, the Genesis creation and worldwide flood stories. As for ultimate questions, however, science, I believed, was simply a different mode of inquiry from religion. I still adhere to this view as a general matter, but Dawkins convinced me that science can be relevant in determining the probabilities of religious answers to ultimate questions.[40]

Of course, faith may lead folks to reject these probabilities, and unlike Dawkins, I see no reason to assume that faith has nothing to offer in the absence of scientific proof (so long as faith is not used to distort or mold science in disingenuous ways). Significantly, at this point in scientific history, Dawkins's faith in science's ability to demonstrate an exceedingly low probability of God's existence must rely on assumptions and presumptions based on lack of evidence.[41] Here he enters the territory of the designers.[42]

The designers rely on so-called gaps in evolutionary biology to suggest that something must fill the gaps, and of course, it is Big D.[43] These gaps are really just the normal areas in scientific development where evidence is currently lacking or there are not enough data to determine the scientific implications of the scientific evidence.[44] Such gaps are necessary for science to do its job, and they are frequently filled as scientific knowledge progresses and new gaps are created as new questions arise.[45] In fact, not having answers assumed up front, and waiting until scientific evidence is available before jumping to conclusions, simply demonstrates that science is doing its job.[46]

The designers see gaps as a weakness – or at the very least a marketing tool for design – and they simply assume that the lack of evidence for a given question means that something supernatural must be at work.[47] Of course, this is because they presume from the start that something supernatural is at work.[48] They never question their ultimate hypothesis and would likely ignore or distort for marketing purposes any evidence that disagrees with their ultimate hypothesis.[49]

Dawkins seems to make the opposite assumption, that is, that lack of evidence – in this case, in science as a whole, not just gaps – demonstrates that there is nothing supernatural or even natural that would equate to God.[50] Just as gaps in evolutionary theory demonstrate nothing about a designer because they simply demonstrate a lack of current evidence on a given point, the lack of proof for God in broader scientific evidence (unless one makes assumptions) is just that – a lack of current evidence. One cannot assume that something exists or that nothing exists based on a lack of evidence. All a lack of evidence demonstrates is that there is a lack of evidence. Perhaps a lack of evidence could somehow support the sort of probabilities that Dawkins addresses, but even then, some sort of testable and falsifiable positive evidence would need to serve as a scientific baseline.[51] As will be seen, no such baseline exists given current scientific knowledge.

Thus, Dawkins enters the proof game (or in his case, the disproof game) on an ironically similar footing to the designers.[52] Dawkins does, however, present a strong positive argument. As was explained earlier, this argument suggests that God would have had to evolve to be complex enough to do any of the things that God did in a variety of faith traditions, and therefore the probability of God's existence is exceedingly low.[53] The argument, however, assumes that nothing supernatural is likely to exist, and as was explained earlier, it assumes that the same physical laws apply not only throughout our universe but throughout all universes, thereby basically dismissing the possibility that what we call God is actually a natural phenomenon that exists under different physical laws or evolved in another universe through the process of evolution.[54] This, in turn, is based on a very Christocentric and Eurocentric view of God.[55] The question whether the God of the Bible, the Koran, or any number of other religious books (which may include polytheistic gods) exists is not the same question as whether a deity – whether a hands-off deity (a Deist sort of God) or a hands-on sort of deity – exists. One need not believe every story in a religious book (or even most of them) to believe that a deity is, or has been, at work.[56] Dawkins focuses so much on more Orthodox religious traditions in creating his baseline argument that his argument is inherently likely to lead exactly where he takes it.

Does the "God must have evolved" argument decrease significantly the probability that God exists? It is certainly an argument worth considering. If we knew that the same scientific laws applied throughout our universe and all other universes, it would be hard to argue with the logic. After all, even if our God evolved in another universe, the question would remain whether that universe had a God and so on. Of course, while there are some educated guesses, we know no such thing. Science is amazing and opens up new doors every day. As Dawkins wrote, it is beautiful, but it has not yet answered the questions mentioned earlier. Even if it did – that is, the evidence was overwhelming that the same physical laws applied throughout our universe and any others – faith may dictate that something supernatural exists, and from a theistic evolutionist standpoint, natural selection may be a mechanism it uses.

This is a question of faith, and the answer can coexist with scientific reality.[57] One can argue that Dawkins's current educated guess will at some point be proven. Even then, however, faith in God will remain for many people because they can, and most likely will, argue that God exists outside of those physical laws and that science is unequipped to answer all the questions in that metaphysical realm.[58]

Lest my point be misunderstood here, it is clear that if the probabilities skew heavily against the existence of God, Dawkins will have won the scientific argument. To argue otherwise is to prostrate science at the foot of religion, which, as I have written elsewhere in this book, is immensely unscientific and misguided. The only way for religion to respond under these circumstances that is not completely self-defeating is to acknowledge that belief in God continues not because God is scientifically likely or even possible but because science and religion are different modes of inquiry and science is not the end of the discussion for people of faith. Those making these arguments, however, would have to accept that their views are scientifically invalid and based strictly on faith. Such a scenario would be especially hard for theistic evolutionists to grapple with, but in the end, it remains a question of baselines.

As Kenneth Miller has explained, theistic evolutionists start out with a baseline that science and religion can coexist without conflicting because faith need not be so rigid as to deny scientific evidence.[59] Dawkins and others begin with a baseline that science is the only valid mode of inquiry and that absent scientific evidence, we are left only with bunk.[60] ID advocates and creation scientists begin with a baseline that there absolutely is something supernatural out there and that science must yield or bend to this presupposition.[61]

Neither of the first two baselines reject scientific evidence because it is inconsistent with a belief in the supernatural. The theistic evolutionist position acknowledges, at least implicitly, that if theism and science ever conflict, the former will not dictate a watering down of the latter but rather the former may be in a different realm from the latter.[62] The pure science position clearly would not reject scientific proof in the absence of anything other than more convincing scientific proof.[63]

It is only the third position, taken on by ID advocates and creation scientists, that seeks to change the ground rules for what counts as science and would accept as valid only evidence that does not preclude a supernatural creator or designer.[64] As was explained in earlier chapters, this is a result of the religious apologetics at play within the ID movement. It is also a result of failing to accept that in the face of overwhelming scientific proof, even the best intended arguments will not stand up to scientific scrutiny if they are simply the result of an apologist marketing scheme. ID errs precisely because it seeks to play the scientific proof game and is unequipped to succeed. Theistic evolution and traditional creationism have no such problem because neither seeks to bend science to prove faith assertions.

Yet if science demonstrates that God is exceedingly improbable, theistic evolution would be in a quandary unique among the three positions set forth previously. As theistic evolutionists, we accept scientific proof and science generally. Where theological positions conflict with science, such as the flood story, science prevails. The key is that in the end, this does not preclude God because the scientific evidence simply proves we evolved. It does not disprove that natural selection is a mechanism that a deity could have used, and in fact, given its efficiency, over time, it may be a particularly good tool for a deity to use.[65] But in the scenario at hand, it is scientific proof that demonstrates that God is exceptionally unlikely.

How could those of us who accept science and faith respond? One response would be to argue that science and religion are separate modes of inquiry and that we can accept that science has proven that God is unlikely while still having faith that God exists based on our metaphysical beliefs. Thus, in the scientific realm – the realm of the natural world – God has been disproved or proven to be exceptionally unlikely, but the faith realm allows for possibilities outside the natural world. The scientific evidence is fully accepted even on the metaphysical questions but only to the extent that science can determine such things.

Others might view this as a cop-out and simply accept the scientific data as proof that God does not exist or most likely does not. The seeming conflict is resolved in favor of the scientific evidence. Atheism or nearly atheist agnosticism would be the only path remaining.

Finally, some may decide to reject the scientific evidence to support their faith positions. This approach would remove those adhering to it from the realm of theistic evolution. The first two approaches can be consistent with the concepts underlying theistic evolution because the first accepts the scientific proof but acknowledges a nonscientific belief that need not lead one to reject what science has shown, while the second leads to the abandonment of religion when the scientific proof becomes too overwhelming. This latter possibility is always lurking for those of us who adhere to theistic evolution, but that is simply because we accept the validity of science.[66] Whether one chooses the first or second response would be a matter of how one views the relation between the physical world and metaphysical possibilities. Either way, the physical proof would not be rejected in the scientific realm.[67] The third approach rejects science in favor of religion and would likely lead to creationism, albeit perhaps in a less biblically literalist form.

III. Dawkins and Intelligent Design: The Bulldog and the Imaginary Mousetrap

It seems almost unfair to pit a leading scientific thinker like Dawkins against the ID movement. As we have already seen, the ID movement does not fair well when pitted against mainstream science, leading theologians, scientific philosophers, or historians (scientific or religious).[68] Yet the battle between Dawkins and ID is a wonderful microcosm of the scientific arguments against ID. This is because Dawkins directly raises the battle lines between positivist scientific proof and the claims of religious apologists to the mantle of science.[69] Dawkins's query to ID advocates is relatively straightforward. It is the one asked by many scientists, theologians, and philosophers of science. The query is simply, show me. Do not bullshit me, but show me as someone who does not already agree with you that you have some scientifically testable and falsifiable proof of your ultimate hypothesis that is not explained by natural phenomena.[70]

The silence in response to this question is almost deafening. Of course, ID advocates chatter quite a bit to respond, but the answers are based in deductive reasoning or marketing ploys, and the response to the actual question – a question that can only be responded to adequately with scientific proof or acknowledgment that the answers lie beyond science – is silence. Why is this?

The reasons relate to the difference between mainstream scientists and religious apologists. Dawkins asks his question in earnest, seeking testable proof.[71] It is not illusion to him or just a ploy. He honestly would consider any real scientific proof offered.[72] To qualify, however, such proof would need to meet the rigors of science.[73] Simply assuming that there is a designer and trying to show this deductively, even through what ID advocates argue to be evidence, is useless to Dawkins. If you never try to falsify your central assumption, and you cannot test it in any way that does not rely on bold-faced assumptions, you simply cannot adequately respond to his questions.

From the ID perspective, there can be no honest answer. To try to falsify the designer hypothesis would – as we have seen – lead to numerous results that demonstrate natural explanations for virtually everything ID advocates assert.[74] They simply cannot use the scientific method to address their central claims because if they really were to adhere to that method, the answers would defeat the ID movement. An obvious response would be one asserted by some theistic evolutionists, which is that ultimate questions currently lay outside of the bailiwick of science. Of course, this, too, is self-defeating for ID advocates. Theirs is a movement based in religious apologetics, and their main apologist strategy is to wedge God into science by claiming the mantle of science.[75] To acknowledge that their core assumptions lay outside of science would undermine ID's reason for existence. ID advocates' only option is to change the ground rules for what counts as science, but as we saw in earlier chapters, they are basically unequipped to do so philosophically, and doing so would create more problems than courts or society are likely to tolerate. Moreover, changing the ground rules would not answer Dawkins's question; rather it would simply avoid his question and its implications for ID.

Dawkins has been called "Darwin's new bulldog" and "Darwin's Rottweiler,"[76] but in the context of the ID debate, he need only be Darwin's old Chihuahua to make his point. As we have already seen, ID relies heavily on imaginary mousetraps whose parts can serve no other purpose than trapping mice,[77] irreducible complexity (a rehash of the watchmaker analogy),[78] and complexity by design, which does not work unless you presume the outcome and disregard numerous alternatives.[79] Dawkins need not make any staggering scientific arguments to win this battle. He need not attack like a bulldog. He simply needs to be there to point out that science has answered these questions in the scientific realm. From there the dialogue can be taken up by philosophers, lawyers, theologians, and theistic evolutionists to demonstrate why redefining science is not likely to help ID in any practical manner and why ID apologetics are part of an ancient apologist movement that is no longer considered within the realm of science.

Ironically, where Dawkins has taken it on himself to function more like a bulldog or Rottweiler is in the battle against the existence of God or at least the probability that God exists. Here he is not Darwin's bulldog or Rottweiler at all. One might argue that he is "Huxley's new bulldog" or "Huxley's Rottweiler" in this context because his writing is learned but aimed at and accessible to a broad audience and because he focuses quite a bit on metaphysical questions.[80] In the end, however, his arguments for evolution are more persuasive than his arguments against the probability of God. To the extent he links his metaphysical arguments to his scientific arguments, he has three target audiences outside of those who already agree with him: (1) theistic evolutionists, (2) creationists (including those who believe in creation or ID but have never really thought about the alternatives), and (3) those with no clear commitments on the links between science and ultimate questions.

For the first group, his argument for evolution is unnecessary because theistic evolutionists already agree with him – although as an aside, as a theistic evolutionist, I do appreciate Dawkins's arguments for evolution for their lucidity and accessibility. They do not persuade me of anything because they do not need to, but I find they make me more articulate in discussing evolution. Yet it his argument against the existence of God that seeks to persuade theistic evolutionists. Perhaps it has succeeded for some, but as earlier parts of this chapter have explained, it raises more questions than it answers for others.

Those who are strongly committed to the creationist or ID cause are unlikely to be persuaded by reasoned scientific argument and even less likely to be persuaded by less well reasoned metaphysical argument – at least to the extent that those metaphysical arguments do not favor their precommitments. It is possible that those less strongly committed to creationism might be persuaded by Dawkins's well-crafted and scientifically impeccable arguments for evolution. As others have suggested, however, his metaphysical posturing may take away with one hand the potential he has created with the other.[81] Thus Dawkins's eloquent arguments for science generally, and evolution specifically, might create enough cognitive dissonance to persuade a weakly committed creationist that theistic evolution, or even atheism, is a viable alternative, but his arguments regarding the metaphysical implications of evolution may raise the cognitive bar too high to be overcome.[82]

Dawkins acknowledges this. From his perspective, the improbability of the *theistic* in *theistic evolution* makes theism generally an unpalatable alternative to what he sees as the necessary atheistic implications of science.[83] Therefore, for Dawkins, losing possible converts to science because he disregards the possibility of theistic evolution is not a real loss because to prevent the loss, he would have to hypocritically disregard the path to which he believes science inevitably leads. The fact that he is unwilling to do so should surprise no one, and it would be disappointing if he did because one can agree or disagree with Richard Dawkins about metaphysical questions; however, he is not a hypocrite.

It is the third group at which Dawkins's arguments are best aimed, and thus it is for this group that honestly confronting the weakness in his metaphysical arguments is most useful. One may read Dawkins's work and that of theistic evolutionists and be more persuaded by Dawkins, but my goal here is to demonstrate that one must make some presumptions to do so. There is nothing wrong with making presumptions when the evidence runs out, so long as one acknowledges that one is doing so, but one must be aware of the presumptions underlying one's theory before being able to acknowledge them.

Whether God exists is a question that has been asked for millennia. Throughout most of that time, it was answered by religious entities or people who believed they had a clear answer. Their argument, however, was circular. Essentially the argument was, "We believe these texts or ideas, these texts and ideas demonstrate there is a God, therefore there is a God." Some people then sought evidence for God in the natural world and found it everywhere in nature's complexity. Of course, they found this evidence for creation or design because they already believed in creation or design. If one was not bound to a precommitment that God created things, one might not come to the same conclusion.

As I have argued elsewhere in this book, there is nothing inherently wrong with such faith, and in fact I share it so long as it is not used to justify harming others or to distort science. Yet those of us who adhere to such faith, myself included, have to be honest that our precommitments lead us there. It does not mean that those precommitments are wrong, just that they exist and that ignoring them is to dummy down the debate.

Dawkins, too, has precommitments. Some are based purely in science and a belief in science's ability to explain phenomena in the natural world.[84] Theistic evolutionists also share this precommitment.[85] As precommitments go, it is a particularly good one because there is a great deal of evidence to support it, which is one of the reasons that even scientific relativism does not help ID advocates. The scientific method actually can prove things about the natural world and repeat the proof again and again. Dawkins's other precommitments, however, affect his views of God. This does not mean he is wrong. It simply means that until there is better evidence one way or the other, he may be more likely to devalue the probability of God based on extant scientific knowledge.

For those in the third group, the likelihood of being persuaded by Dawkins's arguments on metaphysical questions would depend on where their precommitments lie on the scale from belief in God to nonbelief and on Dawkins's persuasive ability as a writer and thinker, which, as we have seen, is considerable. The problem is that Dawkins's lucid writing and place in the world may create a heuristic that avoids the underlying metaphysical issues. Dawkins's arguments become a proxy for grappling with the question of what science actually demonstrates. Most people lack training in theology and the hard sciences, so this is not unlikely.

My point is simply that like religion, which must rely on precommitments to garner support among the third group, Dawkins must also rely on precommitments, and in the end, those in the third group should not confuse his excellent scientific arguments for evolution with his less decisive arguments against God. Whether religion and science are different modes of inquiry and whether religion could survive in a truly scientific realm are questions open to debate. There is nothing wrong with siding with Dawkins in this debate, but it should not be forgotten that, as the work of Kenneth Miller and others suggests, a debate remains on ultimate questions. This, however, does not mean that there is a serious debate over what science has demonstrated about natural selection. Based on the evidence, evolution is a demonstrated scientific phenomenon, and ID remains a marketing strategy.

6

No Intelligence Allowed or No Intelligible Science?

I. The Allegation

"I am being discriminated against in the academic environment because I don't buy into the prevailing theory or methodology in my field and have been engaged in research that calls the prevailing approach deeply into question! Moreover, the established hierarchy in my field is retaliating against me generally so no one will take my work seriously. They are doing this because my work threatens their theories." Claims like this would and should get the attention of anyone concerned about fundamental fairness and the academic enterprise.

Intelligent design (ID) proponents make exactly this sort of claim. In fact, these allegations of discrimination and retaliation by ID proponents were the subject of a documentary by Ben Stein titled *Expelled: No Intelligence Allowed*.[1] Of course, these allegations make several assumptions: (1) that the individual's research is actually in the field addressed; (2) that the reason the person is being discriminated against is because the person disagrees with the prevailing theory and methodology, which means that anyone who so disagreed would receive similar treatment regardless of the merits of the research produced; and (3) that the reason for the alleged retaliation is the perceived threat to the prevailing theory or theories (which would be retaliatory) rather than the merit of the research the person has produced (which would be based on the merits of the research, not retaliation).

If these assumptions are made, the argument is compelling. From a marketing perspective, this a brilliant strategy: claim victim status and accuse the establishment of discrimination. In America, at least, the underdog has a certain amount of appeal, and the intellectual establishment is not always viewed in a positive light. By exploiting these factors and the natural outrage people feel when hearing that someone was discriminated against or retaliated against because of his or her views, the ID movement can – and has – gained some leverage in the public consciousness. These are effective rhetorical arguments regardless of their factual merit. They may even gain the support of politicians, which certainly can help an upstart movement.[2] But what about the substance of the arguments?

Let us rephrase the preceding claims, while addressing the underlying assumptions. "I am being excluded from the academic sciences because I don't buy into the prevailing theory in biology or scientific methodology generally. My research has been rejected because I invoke supernatural causation and refuse to question the existence of that causation. Moreover, the established hierarchy in my field will not take my work seriously because they say I have not engaged in any convincing research, but they tell me that if I can do so, they would consider my ideas. I have engaged in research and they simply reject it. They say it is because it is not persuasive and because there is a huge amount of research by numerous researchers and scholars that refutes my research and that my research is filled with unquestioned assumptions, but I know it's because they disagree with my views." Doesn't pack quite the same punch, does it?

As with the so-called scientific claims made by ID proponents, the claims of so-called discrimination made by ID proponents make a great deal of sense as a marketing strategy aimed at people outside the scientific community but little sense within the scientific community. There is little doubt that if ID advocates could produce adequate research to persuade scientists, even the most hardened materialist scientists would eventually come to accept it.[3] Of course, even ID advocates have acknowledged that for ID to gain acceptance as science, the ground rules for what is considered science would have to change, and assumptions about the supernatural would have to be considered as part of science.[4] While I may be explaining this less charitably and far more blatantly than ID advocates would, that is part of the power of words – a power ID advocates understand well and harness brilliantly.

Of course, one's ability to market one's ideas to the general public and a self-selected audience of those who already agree is a far cry from a persuasive scientific or legal argument. It is, however, consistent with the long-standing approach of Christian apologetics on these issues: convince the people and the politicians rather than the scientists. Yet these claims of discrimination only exist if science is being discriminated against. No one would question a science department for excluding basket weaving, art history, constitutional law, or yes, even alchemy from support or for excluding those engaged in research in these fields from the science department or science generally. Would excluding these folks be discrimination in the obviously negative sense of the word that we have come to accept and that ID proponents intend?

Is it discrimination to exclude a subject from a discipline if it cannot meet the requirements of that discipline or provide persuasive evidence to change enough minds in the discipline to alter the requirements? This is in a sense a basic question of categories. Is it discrimination to say that asparagus is not a fruit but rather a vegetable? It is a form of discrimination in that one is saying that asparagus is not a fruit and therefore is kept out of that category. If it were possible to provide persuasive evidence that asparagus is indeed a fruit, this categorical exclusion might end, but for now, asparagus is not a fruit.

This is, of course, a very basic example. The analogy to ID, however, is obvious. Scientists view ID as a philosophy or religious concept (if they view it at all) but not as science. It does not belong in that category any more than astrology does. We could change the category "science," but to do that in any coherent way, we would need to allow astrology and the rest into science along with ID. As Chapters 2 and 3 explained, this would lead to absurd practical and legal results.

Is it discrimination to exclude nonscience from science? It is in a sense. It means that the category science is limited and that some things – astrology, English literature, Continental philosophy, biblical hermeneutics – are not science. Yet this is true of any field or any category. If science is to have meaning as a concept, no matter how broadly we define it, some things will be excluded from science. The only way around this sort of discrimination would be to consider everything to be science.

So how might anything that is not currently considered science become part of science? Evidence! Provide adequate research to persuade scientists that your approach has scientific merit.[5] ID has failed to do this. It must redefine the category to be included. If such a redefinition does not occur, however, is excluding ID from science and ID advocates from scientific discourse discrimination in any invidious form?

ID advocates might respond, "But the definition of science imposed on us is based on naturalism and materialism, and these ideas themselves are subject to question, yet scientists don't question them." In other words, ID advocates argue that the deck is stacked against them and that the deck itself is discriminatory. The response from the scientific community is that science is all about explaining natural processes and the natural world through demonstrable evidence, and if supernatural causation is ever to become part of science, it would have to be demonstrated through the scientific method. Theistic evolutionists might add that this does not preclude belief in the supernatural, or even belief in a designer, but rather demonstrates a boundary between science and nonscience.

Therefore ID advocates have two choices: (1) play and win the proof game or (2) redefine science. As we have seen, ID advocates have played the proof game as apologist creationists did before them, and like their predecessors, they have not won. Moreover, to redefine science, ID would have to accept scientific relativism with all its implications for practical science and for the underlying tenets of ID. Unless or until ID can meet one of these challenges, any claims of invidious discrimination are factually absurd. There is no more discrimination by science against ID proponents than against alchemists and astrologers. As with many other claims by ID advocates, these claims of discrimination are a savvy marketing strategy, but when the underlying facts are set forth, there is no substance to the claim that ID advocates are the victims of unfair discrimination. Still, it is worth looking at the legal implications of these claims to determine if there is any "there" there, even when an ID advocate is employed as a scientist and claims that his or her ID research is science.

II. The Response in the College/University Setting

A number of the "experts" cited by ID proponents are professors at public universities.[6] Though a number of these experts are in philosophy or religion departments, a few work in science departments, including biology.[7] The question naturally arises whether these universities could preclude professors from teaching or researching ID as faculty members in a science department. Relatedly, there is the question of whether allowing these individuals to teach ID in the science curriculum would be an establishment of religion. Underlying these issues is the proof game because if ID is not science, university officials have much wider latitude.

As will be seen in this chapter, universities may determine that ID is not science and thus preclude professors from teaching it in science departments. Such a determination has significant doctrinal and philosophical implications. Science departments may also consider research by ID proponents under their general tenure policies, which usually require publication in peer-reviewed journals and the evaluation of a tenure candidate's scholarship by outside evaluators in the relevant scientific field (i.e., biology, chemistry, geology, astronomy, etc.). If the scholarship is deemed inadequate by such experts, the department, college, and/or university may take whatever action they would normally take where evaluators find scholarship lacking in quality or depth. This does not preclude the possibility that such work may be appropriate in a religion or philosophy department.

Moreover, this chapter will explain why teaching ID in a public university science department may pose establishment clause problems depending on the specific facts involved. Questions of academic freedom and free speech generally are highly relevant in addressing these issues. Moreover, these issues are directly relevant to the ID movement's marketing strategy. Rather than addressing the reasons ID may be excluded from science departments, or acknowledging that it may be addressed in religion or philosophy departments, ID advocates paint these exclusions as cases of invidious discrimination and themselves as victims of an organized campaign of persecution and retaliation. As will be seen, there is much more to the situation than the ID advocates suggest. Still, while there are valid reasons to question ID advocates' claims of discrimination and persecution, it is hard to deny that painting these situations as examples of discrimination is a brilliant marketing strategy. It also, of course, raises a number of legal questions we will address here.

A. Academic freedom, curricular needs, and disciplinary boundaries

The ability of public universities to preclude the teaching of ID in science classes may seem clear to some. Similarly, the ability of a science department at a public university to deny research support for or tenure to those whose research is primarily focused on ID may seem equally clear. Yet the questions raised by these scenarios are not as easy to answer as it may appear at first glance. To answer these questions, the question of whether ID is science again takes center stage. If it were science, it would be hard to exclude it from the curriculum, at least in upper-level electives, although the courts do give significant deference to university curricular decisions.[8] Because, as we have seen, it is not science in any sense recognized by the scientific community, the deference given by courts to departmental and university curricular decisions enables university officials to keep ID out of science courses.[9]

This still leaves the question of research on ID, and here the courts tend to suggest more deference to the academic freedom of the faculty member.[10] Of course, even this academic freedom is not boundless.[11] For example, one would not expect that a geology department would have to credit, fund, or otherwise support research arguing that the Earth is flat. Nor would an astronomy department have to credit or support research attempting to prove (but not disprove) that our solar system is the center of the universe. Yet as a theoretical matter, one may argue that in precluding such work, one is favoring a particular paradigm for science over other possible paradigms.[12] This argument raises significant philosophical questions,[13] but as Chapter 2 explained, it does not help ID proponents despite its superficial appeal.[14]

B. Intelligent design redux: Religion, science, or religious science?

Is ID science, religion, or both? Proponents of ID argue that it is science and that at least on its face, it is not religion.[15] Yet when probed further about who the intelligent designer might be, the answer is generally God or some sort of unnamed supernatural force.[16] Does this make ID religion? Does it preclude it from being science? As we have seen, we first must determine whether these claims are meaningful given the history and tenets of ID. When this is done, we see that ID closely resembles a number of other movements focused on religious apologetics. ID is a well-supported and rhetorically sophisticated marketing plan, but unless scientific relativism becomes the scientific norm for practicing scientists and scientific organizations, ID's claims to the mantle of science are just as shaky as creation science's claims.

Whether Big D, the Flying Spaghetti Monster, or God, the designer is a supernatural entity, and ID, like creation science, pits itself against naturalism. As we saw in Chapter 3, the one court that has explored the issue in depth held that ID is not science and that it is a religiously grounded approach.[17] This holding is consistent with the evidence that was before the court,[18] and it is also consistent with the generally accepted definition of science and the view of the mainstream scientific community.

Recall that the *Kitzmiller* court heard testimony from leading philosophers of science,[19] biologists,[20] and ID proponents.[21] After hearing all this testimony and evaluating a great deal of other evidence, the court held that ID is not science and that it is a religiously grounded approach.[22] The court's holding that ID is a religiously based approach and not a scientific theory was an important aspect of the decision for a number of reasons but especially because the U.S. Supreme Court had already held in *Edwards v. Aguillard*[23] that religiously based theories of creation could not be taught in public school science classes without running afoul of the establishment clause.[24]

Theologians and philosophers of science testified that ID is not a scientific theory and that it is religious in nature. For example, the court stated the following:

> We initially note that John Haught, a theologian who testified as an expert witness for Plaintiffs and who has written extensively on the subject of evolution and religion, succinctly explained to the Court that the argument for ID is not a new scientific argument, but is rather an old religious argument for the existence of God. He traced this argument back to at least Thomas Aquinas in the 13th century, who framed the argument as a syllogism: Wherever complex design exists, there must have been a designer; nature is complex; therefore nature must have had an intelligent designer. Dr. Haught testified that Aquinas was explicit

> that this intelligent designer "everyone understands to be God." The syllogism described by Dr. Haught is essentially the same argument for ID as presented by defense expert witnesses Professors Behe and Minnich who employ the phrase "purposeful arrangement of parts."
>
> Dr. Haught testified that this argument for the existence of God was advanced early in the nineteenth century by Reverend Paley and defense expert witnesses Behe and Minnich admitted that their argument for ID based on the "purposeful arrangement of parts" is the same one that Paley made for design. The only apparent difference between the argument made by Paley and the argument for ID, as expressed by defense expert witnesses Behe and Minnich, is that ID's "official position" does not acknowledge that the designer is God. However, as Dr. Haught testified, anyone familiar with Western religious thought would immediately make the association that the tactically unnamed designer is God.[25]

Another example follows:

> Robert Pennock, Plaintiffs' expert in the philosophy of science, concurred with Professor Haught and concluded that because its basic proposition is that the features of the natural world are produced by a transcendent, immaterial, non-natural being, ID is a religious proposition regardless of whether that religious proposition is given a recognized religious label. It is notable that not one defense expert was able to explain how the supernatural action suggested by ID could be anything other than an inherently religious proposition.[26]

As Chapter 2 explained, the arguments made by ID advocates for design go back even further than Aquinas. They are at least 2,000 years old and can be traced back at least to Plato. Significantly, throughout that 2,000-year history, none of the design advocates from Plato to Cicero, Aquinas to Paley, pretended that the designer was anything other than a deity. Design has always been a form of religious apologetics, and throughout most of its history, no one saw any reason to hide this fact.

One of the most problematic pieces of evidence for ID proponents is the Wedge Document,[27] which was discussed in Chapter 1. The Wedge Document makes clear that ID is essentially a way to get the Christian view of creation back into public school science classrooms after the defeat of creation science in *Edwards v. Aguillard*.[28] The document reads more like a marketing strategy for Christian apologists interested in overthrowing evolution than the basis for a scientific school of thought, and this is, of course, consistent with the unacknowledged religious apologetics underlying the ID movement. As the *Kitzmiller* court explained,

> Dramatic evidence of ID's religious nature and aspirations is found in what is referred to as the "Wedge Document." The Wedge Document, developed by the Discovery Institute's Center for Renewal of Science and Culture (hereinafter "CRSC"), represents from an institutional standpoint, the IDM's goals and objectives, much as writings from the Institute for Creation Research did for the earlier

> creation-science movement. . . . The Wedge Document states in its "Five Year Strategic Plan Summary" that the IDM's goal is to replace science as currently practiced with "theistic and Christian science." As posited in the Wedge Document, the IDM's "Governing Goals" are to "defeat scientific materialism and its destructive moral, cultural, and political legacies" and "to replace materialistic explanations with the theistic understanding that nature and human beings are created by God." The CSRC expressly announces, in the Wedge Document, a program of Christian apologetics to promote ID. A careful review of the Wedge Document's goals and language throughout the document reveals cultural and religious goals, as opposed to scientific ones. ID aspires to change the ground rules of science to make room for religion, specifically, beliefs consonant with a particular version of Christianity.[29]

If ID is not science as accepted by the broader scientific community, and it is a religiously based approach, as found by the court in *Kitzmiller* and as reflected in many of the ID movement's own documents and statements, what arguments remain for ID advocates if a college or university science department excludes ID from teaching or research? After all, if it is not science, it need not be taught or recognized by science departments at universities[30] and in fact, as explained later, could be kept out of the science classroom.[31] As will be seen, the best remaining argument for ID advocates ultimately works against ID advocates' claims that ID is science.

We have already seen that the ID movement's best – and probably only – argument to escape this conundrum is that the definition of science accepted by mainstream science is too narrow. To make this argument, ID advocates must conceptually rely on the notion of multiple scientific paradigms,[32] that is, on scientific relativism. Yet Chapter 2 explains why such arguments prove a bit too much for ID advocates. After all, under this approach, the scientific method would only be a tool of the currently prevailing paradigm for science, but it would not be necessary for all scientific paradigms.

Of course, this argument would allow alchemy, UFOlogy, and astrology to be considered science as well. One would hardly expect that chemistry departments would accept alchemy as an appropriate teaching or research field, nor would one expect an astronomy department to credit teaching or research focused on astrology (one can imagine the Dionne Warwick space telescope instead of the Hubble).

Of course, as Thomas Kuhn demonstrated, ID would not likely be successful – in the sense of being accepted as science by anyone other than those who already believe in or agree with its underlying presumptions – because it has not and will not likely be accepted by any community of credible scientists even if it were considered a scientific paradigm.[33] Chapter 2 explained that even if, at a metaphysical level, there are multiple scientific paradigms, there remains a practical substantive boundary for what can be called science, even if in narrower terms, that boundary may shift.[34]

Astrology, ID, and the belief that the Earth is the center of the universe are all precluded from science because they do not use the tools, quantitative analysis, or methodology of science in regard to their ultimate hypothesis.[35] Still, in arguing that he or she is being discriminated against or denied academic freedom, an ID advocate might ask how we know that a given paradigm is *the* paradigm for a given science unless there is some superparadigm that allows us to choose between competing paradigms, and there is no such superparadigm.[36]

This is an interesting argument from relativist metaphysics, but it hardly helps ID when one thinks it through as a practical matter – although it does add some cache to the ID movement's marketing plan. As the father of the scientific paradigms argument, Thomas Kuhn explains, and as was discussed in Chapter 2, despite the metaphysical possibility of multiple paradigms there are still criteria for what can be counted as science (i.e., what already is science or might cause a paradigm shift in science),[37] and ID does not meet these criteria.[38] Perhaps more important, when courts address questions of academic freedom within a given discipline, the metaphysical question implicitly gives way to the more basic question of whether the disciplinary boundaries within a given field can be maintained while allowing for robust academic freedom.[39] Of course, since ID cannot demonstrate that it is science within the Kuhnian world of shifting scientific paradigms, the legal questions are even easier to answer.[40]

This is not to say that academic freedom within the sciences is less robust than in other fields. A good example of a highly controversial theory that has gained a good deal of acceptance while also garnering a good amount of skepticism is the field of string theory in physics.[41] Of course, one of the reasons the theory is so controversial is that it is hard to falsify based on real-world observation (it is primarily a set of mathematical models supported by *some* real-world research).[42] Of course, string theorists do not use this to avoid the scientific method; rather, they have endeavored to analyze (i.e., prove or disprove) their theories using more and more sophisticated experiments and equipment.[43] One of the most recent examples is the CERN Large Hadron Collider that was recently completed in Switzerland,[44] which is even larger than the previous world champion atom smasher at Fermi National Accelerator Laboratory in the United States.[45]

ID advocates, of course, do no such thing.[46] There is no attempt to prove *or falsify* the existence of the intelligent designer through scientific experiments. There is an attempt to prove the designer's existence using concepts such as irreducible complexity – the notion that some biological phenomena such as the human eye or a bat's ability to fly could not exist based on evolution because if you remove one part, the entire function falls apart – to demonstrate the existence of the designer.[47] Virtually every example ID advocates provide for irreducible complexity is countered by modern science.[48] For example, it is well understood based on the fossil record and through experiments that some biological functions resulted from changes in the function of a specific body part.[49] Thus all parts of complex biological functions need not have evolved initially to serve that function.[50] There are many examples of such complex systems in nature and in the fossil record.[51] Yet ID advocates do not generally acknowledge this research, except to state that it is wrong.[52] They do not attempt to falsify their fundamental principles.[53]

String theory is the ultimate foil for the claims of ID advocates because it has become an accepted paradigm among many theoretical physicists, even as others suggest that the inability to falsify it makes it more philosophy than science.[54] As noted earlier, string theorists do not respond by saying, "We don't need to prove the existence of strings because they are so well demonstrated by nature."[55] They argue that strings are well demonstrated by phenomena in the physical universe,[56] but at the same time they acknowledge that until they can prove their ultimate hypothesis through experimentation, they may be going down a path that will prove to be wrong.[57]

Importantly, the reason string theory became accepted was because of the complex mathematical calculations made by leading theoretical physicists that support it as well as the many real-world phenomena it seems to explain.[58] Without the math, however, no one would have accepted it.[59] Whether it ultimately results in a paradigm that is accepted by the community of physicists will be based on its ability to use the tools of that community to demonstrate and convince others that it is superior to earlier paradigms.[60]

The evolution of string theory itself is practically a page out of Kuhn's work. After some initial promise, it became more marginalized, as quantum mechanics (the study of the very small, such as atoms, in physics) and cosmology (generally connected more with relativity and the study of larger phenomena in the universe), went in other directions.[61] After years of complex mathematical computations, the theory had a rebirth of sorts and became quite popular.[62] It was said to merge things on the quantum level and relativity; that is, it merged gravity with the three fundamental forces of quantum mechanics.[63] This led to an explosion of work in the area that created several mathematical models that seemed inconsistent with each other.[64] This was, of course, problematic if the theory were actually to unify relativity and quantum mechanics. It was also problematic because these models required 10 space-time dimensions instead of the generally accepted 4.[65] Yet the calculations made sense, and extra dimensions, particularly at the quantum level, explained a number of natural phenomena.[66]

Then Professor Ed Witten, one of the world's leading theoretical physicists, determined that the various mathematical models could actually gel because they were all reflections of the same phenomenon in various contexts.[67] This led to the revelation that there is an 11th dimension in string theory.[68] The 11th dimension solved a number of problems in the theory and also allowed its broader application at the level of cosmology.[69]

The problem remained, however, that all of this was hard to prove or falsify in the physical world.[70] The mathematical calculations were complex and meticulous, but did they explain anything real? Some physicists argued that unless string theory is falsifiable, it is not science but rather philosophy.[71] String theorists responded that the calculations themselves could be falsified,[72] but moreover, they agreed that until string theory is adequately demonstrated or falsified in the real world through experimentation, the verdict is out on its validity.[73] They then set out to prove or disprove the real-world validity of string theory.[74] Other physicists suggest that this is impossible,[75] but that, of course, is a question of the ability of technology to prove it.[76] If technology cannot prove it, string theory will be seen as a dead end.[77] If it can, then the debate will likely rage on.[78] Thus even the most borderline (between philosophy and science) scientific theories still must prove themselves using the scientific method to prevail.

C. Teaching intelligent design at the university level

There is a significant amount of case law holding that public university officials may insist that professors teach within the stated curriculum.[79] It is equally clear that within the curriculum, professors are accorded a great deal of academic freedom,[80] although there are some limitations.[81] Some of these cases involve professors inserting their religious views into courses unrelated to religion.[82] In the end, courts have held that courses at public universities are so connected with the educational function of these institutions that university officials have a right to enforce "legitimate pedagogical interests" as to the general substance of courses.[83] These interests outweigh any claims of academic freedom asserted by professors,[84] or those claims are said to be invalid when it comes to teaching (at least in the core curriculum).[85] Thus arguments for including ID in the science curriculum based on equal access or formal neutrality in the aid context are inapposite here because there is no public or limited public forum and no facially neutral program of "private choice."[86] This is further backed by the argument that ID is not science because even if there were a limited public forum in this context – and there is not – that forum would be limited to "science" courses in the science curriculum.[87]

At one level, this is a bit disturbing to academics like me. I had thought that academic freedom was quite broad in the classroom as a matter of both law and policy, but reading the cases, it seemed more and more like this is so as a matter of policy but not as a matter of law. Yet the ascendance of ID suggests that there are reasons why the courts have ruled as they have. Most of the cases do not involve garden-variety teaching disputes.[88] They more frequently involve either overt sexualized or profane statements in courses that do not touch on sex or profanity in any way, or they involve the insertion of material that may run contrary to the focus of the courses involved.[89] Many of the cases involve required courses as opposed to electives, and the professors involved frequently teach primarily at the undergraduate level.[90]

The most relevant of these cases for present purposes is *Bishop v. Aronov*, decided by the U.S. Court of Appeals for the 11th Circuit.[91] Bishop was a professor in the Department of Health, Physical Education, and Recreation in the College of Education at the University of Alabama, where he taught exercise physiology.[92] He was also the director of the college's Human Performance Laboratory.[93] The university issued Bishop a letter requiring him to abstain from inserting religious statements in his teaching.[94] The subject matter of Bishop's statements, as attested to by him in an affidavit, included remarks like the following:

> I want to invest my time mainly in people. I personally believe God came to earth in the form of Jesus Christ and he has something to tell us about life which is crucial to success and happiness. Now this is simply my personal belief, understand, and I try to model my life after Christ, who was concerned with people, and I feel that is the wisest thing I can do. You need to recognize as my students that this is my bias and it colors everything I say and do. If that is not your bias, that is fine. You need, however to filter everything I say with that (Christian bias) filter.[95]

Bishop also organized an after-class event for his students and others who were interested, at which he lectured about "Evidences of God in Human Physiology."[96] The session was held shortly before exams, and the university felt this could place pressure on students to attend.[97] Although Bishop utilized a blind grading system, the university did not think he adequately separated the out-of-class lectures from the course itself. He would be able to hold such an event if it was not seen as being associated with the course, but the university saw no such separation between the course and after-class events.[98]

The court held that a university classroom is not a public forum for speech.[99] Thus the university has the right to determine what substance is appropriate in the curricular context so long as it has legitimate pedagogical interests for doing so.[100] This must be done through case-by-case analysis.[101] In *Bishop*, the university had valid concerns over the relevance of the professor's religious statements to a course in exercise physiology.[102] Bishop had the freedom to hold events on his views of God's role in human physiology on campus so long as those events were not connected to his courses.[103] Thus Bishop was not denied the freedom to discuss his religious convictions; rather he was only denied the ability outwardly to do so in the manner that he had in his exercise physiology course.[104]

The key issue was the department, college, and university's right to control the curriculum based on legitimate pedagogical interests.[105] In this case, those interests included concerns about the pedagogical effects of students feeling religiously coerced in a basic physiology course.[106] The notion of legitimate pedagogical interests was taken from a line of cases involving secondary schools,[107] and the court acknowledged that it was borrowing from these secondary school cases – although those cases would have to be adapted to the university setting:[108]

> In an effort to calibrate the Constitution in this case . . . we consider the context: the university classroom during specific in-class time and the visage of the classroom as part of a university course in an after-class meeting. This context also leads us to consider the coercive effect upon students that a professor's speech inherently possesses and that the University may wish to avoid. The University's interest is most obvious when student complaints suggest apparent coercion – even when not intended by the professor.
>
> Second, it follows, we consider the University's position as a public employer which may reasonably restrict speech rights of employees more readily than those of other persons. As a place of schooling with a teaching mission, we consider the University's authority to reasonably control the content of its curriculum, particularly that content imparted during class time. . . .
>
> Last and somewhat countervailing, we consider the strong predilection for academic freedom as an adjunct of the free speech rights of the First Amendment. There are abundant cases which acclaim academic freedom.[109]

In holding that the university did not violate Bishop's free speech rights, the court stated the following:

> Though we are mindful of the invaluable role academic freedom lays in our public schools, particularly at the

> post-secondary level, we do not find support to conclude that academic freedom is an *independent* First Amendment right. And . . . we cannot supplant our discretion for that of the University. Federal judges should not be ersatz deans or educators. In this regard, we trust that the University will serve its own interests as well as those of its professors in pursuit of academic freedom. University officials are undoubtedly aware that quality faculty members will be hard to attract and retain if they are to be shackled in much of what they do.[110]

The court was clear, however, that the university did not have the right to preclude Bishop from engaging in independent research on such topics.[111] This issue will be discussed later.[112]

In the context of ID in the science curriculum, one can glean from the cases that university officials as well as departmental curriculum committees can exclude the teaching of ID if they choose to.[113] The same would be true regarding astrology, alchemy, and so on. In addition to the balancing test from *Bishop*, courts have based such holdings directly on the secondary school cases,[114] that is, determining whether the university's decision is based on legitimate pedagogical concerns and whether the course in question is seen as university speech[115] – which most courts hold it is[116] – and thus distinguishable from cases involving private speech.[117] Other courts base their decisions on the cases involving the free speech rights of teachers for out-of-class speech or the speech rights of government employees generally.[118] These courts generally weigh the interests of the government employee as a private citizen in "commenting on matters of public concern" and the interest of the government as employer in promoting its interests.[119] Still others apply both approaches.[120]

A somewhat analogous scenario might prove helpful. Imagine a law professor who teaches a course on agriculture and law. The professor does not teach anything directly about the law relating to agriculture. Instead the professor teaches seed biology and uses this to demonstrate that the relevant law is "wrong." This is not done by teaching the law in a critical fashion – approaching it from a variety of angles, including its incongruence with seed biology – but rather by using seed biology to attack gaps in the legal framework rather than addressing the framework as a whole. One might expect that after student complaints, a law school might suggest that the professor teach the course in the agriculture school, if that school wants such a course, but not in the law school. The same could be true of courses in ID. A university can insist that such courses be taught in the philosophy or religion departments, if those departments want them, but not in biology courses.[121]

D. Research on intelligent design in public university science departments

Though the case law is quite clear about the right of university officials and faculty committees to affect the substance of certain courses despite academic freedom concerns, it is not so clear regarding the university's role in research. *Bishop* acknowledges that academic freedom is far greater when it comes to research.[122] Yet we know that in hiring, tenure, promotion, and merit increase decisions in the sciences, much depends on the researcher's publication output, ability to get grants from recognized granting sources, and professional reputation among peers. It is also clear that ID advocates are not generally published in mainstream science journals, their work is not highly regarded (if regarded at all) by scientific peers, and their ability to get grants for ID research from mainstream granting institutions is basically nonexistent.[123]

Thus, two questions naturally arise. First, despite the academic freedom to pursue ID research, can science departments choose not to recognize that research as meaningfully aiding the department's research interests (either substantively through grants and publications or reputationally)? Relatedly, could a science department simply exclude ID research from any support or recognition? In other words, could a science department simply decide that ID is not science and therefore that ID research has no place in a science department (or using the name of such a department)? Second, could a science department revoke the tenure of a faculty member who posttenure engages only in ID research and refuses to teach courses that do not include ID?

As will be seen, the answer to the first question is yes, but the answer to the second question is far more complex. There is a vast difference in tenure revocation policies among universities,[124] and if the faculty member is willing to teach his or her assigned courses without focusing on ID (assuming it is not appropriate to the courses) and he or she is willing to engage in appropriate service activities, for example, committee work, for-cause tenure revocation would not, and should not, be allowed under most university policies.[125] It is one thing to deny merit increases to a faculty member because his or her ID research is essentially useless to a department and quite another to revoke a faculty member's tenure because he or she is engaged in so-called junk theory outside his or her teaching and service. Unless a department is willing to revoke tenure for all nonproductive researchers, it seems problematic to revoke tenure because of what is essentially an outside hobby (because it would not count as scholarship).

This seems a logical balance given most university policies.[126] A science department could deny any support for ID research (including the use of the department, college, and university name) and give no credit for it in terms of research productivity, but such a department should not treat an ID person differently than any other nonproductive researcher in terms of tenure revocation. To do so would be a form of viewpoint discrimination that could be actionable under the free speech clause.

The first preceding question involves no special First Amendment analysis given the discussion above. If ID is not science, science departments have no duty to fund it any more than a science department would have a duty to fund a professor's art collection. A department or university would also have the ability to require that its name not be used in connection with the work. For example, if a faculty member wants to engage in a partisan political blog or a blog promoting drug use, a public university would have the right to refuse the faculty member resources for the blog and to require that the university name not be used to promote the blog.[127] This is not required, but the university may do so. The same would be true with ID.

The credit issue is even easier to deal with. Science departments, like other departments, need not support or reward research that does not meet the basic criteria set for such support or reward.[128] If an ID researcher cannot place work in accepted peer-reviewed journals, get grants from (scientifically) credible granting institutions, and/or get favorable peer review from scientists, there is no duty to support the work.[129] It is not science.[130] One would not expect science departments to have to fund UFOlogy research, research on why the Earth is flat, or research examining why the Earth is the center of the universe. The same is true for ID research. Departments could fund such research unless it would pose an establishment clause issue (discussed later), but they need not, and are not likely to fund or reward such research. Of course, ID research might be relevant in religion or philosophy departments.

The question of tenure revocation is quite different. First of all, there is a question of tone. Denying credit to junk science is a refusal to give a carrot to those who do not engage in serious scientific work, but revoking tenure is a punishment. One is based on merit; the other, even if arguably based on merit, is punitive in nature and will most likely be treated by courts as such.[131] The issue is not one that can be addressed in broad terms.

First, there is the fact that various states and public universities have different tenure-revocation policies.[132] Second, there is the practical reality that tenure revocation is a rare occurrence and is not generally based on research alone. As a general matter, for-cause tenure revocation has occurred where there is a complete lack of performance, that is, failure to meet teaching, scholarship, and service duties, as opposed to just one category.[133] Even then, there is generally notice and an opportunity for the faculty member to improve performance as well as general due process rights.[134] Other cases may involve extreme malfeasance by a faculty member such as embezzlement, significant plagiarism, significant criminal conduct, and the like.[135]

Assuming that the faculty member is meeting his or her teaching duties and service requirements (usually involving committee work), tenure revocation would appear more like punishment for the faculty member's religious and or political views. This is not a valid basis on which to revoke tenure. If, on the other hand, a faculty member refuses to teach his or her courses or refuses to teach them without including ID, and that faculty member engages primarily in ID research that does not help, and may hurt, a science department's reputation, tenure revocation would be a possibility (even then, it would depend on university policies, and due process would certainly be required).[136] The reason for the action, however, would be failure to perform even the basic requirements of the job and not the faculty member's belief or research in ID.[137]

The preceding dichotomy between the requirements for refusal to support and reward ID work and revoking tenure makes sense when one views the former as refusal of support and the latter as punishment. A number of cases demonstrate the distinction between the government's ability to grant or deny support to activities performed on behalf of the government and the government's ability to punish private speech by government employees on matters relevant to public discourse.[138] Though ID advocates are likely to argue that denial of research support is punishment, given that the government can choose what speech to fund on its own behalf, the case law is not on their side.[139]

Nor, for that matter, is common sense. Though refusal to fund could be seen as punishment at some level, it is different in kind than revoking rights already given. If the former is considered punishment for speech activity rather than failure to support speech the public institution does not wish to support with its name and money, public universities would arguably be required to fund and/or reward anything and everything that a faculty member claims to be research, including astrology, alchemy, flat-Earth ideas, and Raelian "science." Though a great deal of leeway should be given for research topics in any academic institution, one can see in this argument the long shadow of science departments having to fund research in alchemy, astrology, and flat-Earth ideas. Thus, though great leeway should be given to research topics, that leeway is not limitless, especially in fields with relatively accessible disciplinary boundaries.

E. Intelligent design in public universities and the establishment clause

The primary establishment clause concern regarding ID in science departments at public universities involves teaching ID. Support for research may also be an issue, but as will be seen, teaching of ID poses a far more significant problem under the establishment clause. The *Bishop* court relied in part on the university's justified fear of religious endorsement and coercion when it upheld the university's right to preclude Bishop from teaching a religious approach in his exercise physiology class.[140] An important implication was that this was a general science class and not an upper-level seminar and that it was a science class as opposed to a class in religion or philosophy.[141] These concerns were also echoed in *Kitzmiller*.[142]

Since ID was found not to be science and to be a religiously based approach, the *Kitzmiller* court held that allowing even a disclaimer of evolution based on ID would create endorsement problems.[143] Of course, *Kitzmiller* involved secondary schools and not the college or university setting.[144] *Bishop* engages similar reasoning in the university context, but the establishment clause issue was not directly before the court; rather it was one of the university's reasons for precluding the religious material in the science class.[145] Thus it is worth briefly analyzing the question under the tests generally used in education situations under the establishment clause.

As you may recall from Chapter 3, the legal tests that have been applied in these situations are the endorsement test,[146] the coercion test,[147] and the *Lemon* test (as combined with endorsement analysis or as separate analysis).[148] Teaching ID in public university science classes, as opposed to philosophy or religion courses, does raise significant establishment clause problems, whereas research support for individual researchers (if any credible science department would provide it) does not. The obvious reason for this is the difference between the classroom and scholarly contexts.[149]

When one registers for a course in the science curriculum, one does not expect to have religious positions on creation thrust on oneself.[150] Once one is registered for the course, it may be hard to withdraw for any number of reasons.[151] If the professor imposes his or her religious views on the scientific subject matter of the course or for religious reasons skews his or her teaching to create a false impression that a generally scientific approach is invalid, there are clear problems of endorsement and coercion.[152]

In the endorsement context, the question is whether the purpose or effect of the government actor's actions endorse religion or a particular religion and make religious outsiders feel like outsiders in the political community or religious insiders feel favored.[153] The purpose aspect of this analysis is not as easy at it may appear at first glance. The first question that needs to be asked is, whose purpose are we looking at, the professor's or the university's? If the professor is teaching ID as a valid theory and one looks to the professor's purpose, there would be a strong case that the purpose is to endorse religion.[154] The professor would be teaching as valid a religious idea that is not scientific, and any argument that doing so promotes secular pedagogical purposes in a science classroom is inadequate once ID material is taught as science.[155] If anything, teaching ID as valid science in a science classroom would go against secular pedagogical purposes.[156]

If one looks at the university, however, the question would be whether the university had any purpose to endorse religion in offering the course or allowing it to be taught after complaints. As a general matter, there would appear to be a secular purpose under either circumstance. Certainly offering science courses has a secular purpose, and even if the university is aware of concerns regarding ID, it may allow it to continue based on the university's sense of academic freedom rather than an intent to endorse religion. This issue is of little import, however, because teaching ID as science in a science classroom would violate the effects element of the endorsement test.[157]

As the *Edwards*, *Bishop*, and *Kitzmiller* courts all suggest, the effect of teaching religious approaches to creation in a secular science classroom is to promote or endorse religion.[158] Using the podium of a state university science department to promote a religious approach to origins that has been rejected by the broader scientific community is an endorsement of religion. As the *Bishop* court explained, it could make students feel that they must withstand it (i.e., not object) or have their grades affected,[159] and as the *Kitzmiller* court explained, it can create a false sense of scientific views on central issues in students who do not have a strong grounding in biology, chemistry, and so on.[160]

Certainly those who do not share the supernatural or religious views promoted by ID[161] would be made to feel like outsiders in the political community and those who support that view like insiders.[162] This would be the effect of teaching the material as valid science in a science course – either promoting ID or using its concepts to attack evolution – regardless of whether one views the speaker as the professor or the university.[163] Many students, even college-aged students, may not separate the message from the professor and the university.[164]

Moreover, if the university allows ID teaching to continue after complaints by students, the effect of teaching ID would be even more intense regardless of the university's reasons for not acting to prevent ID in the science curriculum. The preceding analysis would also be applicable under the first two prongs of the *Lemon* test – whether the government entity has a secular purpose and whether the primary effect of the relevant government action is to advance or inhibit religion.[165] The third prong of the *Lemon* test might also be implicated based on the political divisiveness these issues raise, but the current role of this element is unclear, and at any rate, it is unnecessary because the purpose and effect elements are violated.[166]

The establishment clause makes the public university's role in limiting the teaching of ID in science courses mandatory. The likely argument in defense of allowing ID based on free exercise and/or free speech is doomed to failure because the university has the ability to preclude ID from being taught in the science curriculum.[167] Thus there are no free speech or free exercise rights involved because there is no unlimited right to teach whatever one wants regardless of curricular needs or academic merit.[168]

The only exception might be in a situation in which there is evidence that other religious approaches can be taught as science but ID is being singled out for negative treatment. This could potentially violate the free exercise clause. Of course, this would only apply if other religious approaches – such as Scientology or Raelianism – are being taught as science. It would not apply if ID is excluded and no other religious approaches are involved.

Under the preceding analysis, ID would be treated in the science curriculum in the same way that astrology, alchemy, or UFOlogy are treated, except that since the latter ideas are not generally religious in nature, there would be no duty to preclude the material (of course, no science department would likely allow the material to be taught as valid science). The one exception on this latter point might be religious versions of UFOlogy, such as Raelianism,[169] which are not generally part of the "mainstream" UFOlogy community.[170] To the extent such religious motivations are the reason for teaching the material, and assuming a university would consider allowing Raelianism in a science classroom, the same establishment clause analysis would apply to it as applies to ID.[171]

Finally, courts might apply the indirect coercion test to these situations. This test explores whether the government sponsored a religious exercise that obliged the participation of objectors.[172] The teaching of ID as science in a science classroom would certainly oblige the participation of objectors who may not be able to drop the course without significant concerns and, if the course is required for any reason, may not be able to drop it at all.[173] The government certainly sponsors the class, and a state actor is the one promoting the religious approach on state property, using state equipment, while being paid by the state. The only question would be whether teaching ID would be seen as a religious exercise. Given the discussion of ID in earlier chapters and its supernatural assumptions (that just happen to be consistent with some of the religious tenets of Christianity), and the fact that it would be taught to a captive audience in the classroom, it is likely that most courts that confront the issue would find the teaching of ID in a science classroom to be a religious exercise. The age of the students, which can be important under the indirect coercion test,[174] would not preclude application of that test here because even though the students are college aged, as in *Bishop*, they are a captive audience being taught by a state employee.[175] As noted earlier, when one enrolls in a science class, one expects that one will be taught science and that religion, if it is involved at all, will not be taught as science.[176] One might, however, seek such discussion in philosophy of science, general philosophy, religion, or maybe even history of science courses.

F. Discrimination?

In the end, any suggestion that ID advocates have been victims of discrimination because their scientific ideas have been rejected by the academic and scientific establishment, is not factually or legally defensible. Their ideas are not scientific under the standards used by mainstream science. ID advocates are correct, however, that their ideas have been rejected by the academic and scientific establishments, but this is not because scientists have a special hatred for ID; rather it is because scientists expect science to follow the rules of science and ID does not.

Is it possible that some scientists are more hostile to ID than other supernaturally based approaches? I believe this is possible. But two questions must be raised in response to this possibility. The first is, why would there be special hostility toward ID? The second asks whether such hostility is legally relevant discrimination even if it exists.

Many scientists care little about ID and harbor no special feelings toward it one way or the other. It is as irrelevant to them as astrology. Others, however, may be specially bothered by supernatural causation, especially when made as a scientific assertion.[177] Why is this so?

The title of this book might provide a hint. Unlike most supernaturally based approaches, ID, as did creation science before it, sees itself as an alternative to mainstream science and, more important, has aggressively marketed itself as such. It has tried to use the political process and appeals to the general public to win its scientific battles. This is not an appealing approach to scientists who have spent time in the trenches of scientific research. Then there is the matter of ID's "scientific arguments." Most scientists who have heard these arguments, especially those familiar with evolutionary biology, immediately see the flaws and assumptions in the arguments.

Most of the scientific assertions made by, and examples given by, ID advocates have been proven wrong by mainstream science.[178] Moreover, the scientific arguments made by ID advocates fail to question the ultimate existence of the designer. They seek to prove the existence of the designer, which they already presume. This is not a likely way to win mainstream scientists over.

ID advocates have also spent a lot of time attacking evolutionary biology from the shadows, with arguments like gaps and teach the controversy. As the next chapter will explain, these are powerful rhetorical tools, but they are infuriating to those who realize how ridiculous these arguments are – especially the gap argument – from the scientific perspective. It should come as no shock that many scientists find this sort of rhetorical science distasteful and deceitful. Many may also be concerned that acceptance of ID will lead to a dummying down of the public and a backward turn in scientific and technological progress.[179] Of course, most scientists are happy to go about their business without giving ID any serious thought.

Let us assume for the sake of argument that the sort of hostility mentioned above exists and that an ID researcher, or ID generally, is excluded from, or even ostracized from, scientific fora. Is this discrimination in any actionable form? If ID had scientific merit, and ID researchers were able to provide convincing scientific proof for design, exclusion would be unscientific. This, of course, is the nature of science. If someone provides convincing proof for a previously unaccepted theory, it may come to be accepted as science, but it will be put to its proofs. String theory is an excellent example of this. ID has not provided any convincing proof that would give it a place at the table of science.[180]

To determine whether university officials, academic researchers, or scientific organizations and fora have discriminated against ID advocates, the lack of scientific proof supporting ID is the central issue. Assume astrologers are aggressively marketing astrology as science and seeking to convince politicians and the public of astrology's scientific value by using fancy terminology that sounds scientific, but they have not produced any convincing scientific proof. A faculty member in the astronomy department at a major university decides to devote his or her time to astrology, and as a result, the faculty member is denied promotion or a research stipend. Is this discrimination? Assume even that the department chair and faculty are particularly hostile to astrology when astrologers assert that it is science. Assume also, that if the faculty member had been able to publish his or her work in credible journals and bring in grants from credible scientific organizations, there would be no problem. The faculty member has not done so, however. This is not discrimination. It is common sense not to credit nonscience as science.

ID advocates with scientific credentials have been able to avoid some of these problems because they can still do mainstream research and get it published. They may then argue that it somehow supports ID. For example, Michael Behe has done this.[181] The arguments for design need not be given credit by a science department, and in fact, as Kenneth Miller has shown in Behe's case, the underlying research does not support ID.[182]

What is interesting here is the unwritten double standard ID advocates apply when it comes to this alleged discrimination and exclusion. When ID advocates do mainstream scientific work, they are able to get it published in scientific journals. They are not discriminated against by the entire scientific community because they believe in ID. It is only when they make nonscientific arguments, ones that assume the existence of the designer and then set out to prove it through deductive reasoning and inference rather than falsifiable evidence that can be reproduced by others, that scientists fail to give them credit.

In fact, ID advocates regularly tout any mainstream scientific publication by any ID advocate, even when the research does not prove anything one way or the other about design. If the entire scientific universe is conspiring against ID advocates, how do these publications exist in the first place? Or is it simply that scientists and scientific venues will give scientific research a fair shake but not a marketing strategy claiming the mantle of science?

One final note on the discrimination and exclusion issue is in order. None of the preceding means that there will not be isolated cases where scientists do not treat colleagues who profess a belief in ID with the humanity with which we hope all people will be treated. The case of astronomer Guillermo Gonzales might be such a case, but of all the supposed incidents of discrimination and expulsion from the academic community, his is the only one that involves a scientist doing a decent amount of mainstream research and having a number of mainstream publications who was both denied a major job benefit and subject to personal vitriol in the workplace.

As was explained in Chapter 3, if an ID advocate is harassed based on his or her religious commitments and that harassment is severe or pervasive enough to meet the requirements for a hostile work environment based on religion, the ID advocate would have a potentially valid claim of discrimination. This would not be because ID is science but rather because no one, scientist or otherwise, has a right to discriminate based on a protected classification such as race, religion, gender, or national origin. Excluding ID from science is not such discrimination, but attacking someone's religious commitments may be.

7

The Madison Avenue Approach to Law and Science

As the scientific historian Ronald Numbers explained, not many creationists adhered to any view of scientific creationism until the early part of the twentieth century.[1] To the extent that science was inconsistent with creationism, most creationists either rejected the scientific evidence or viewed it as part of God's plan.[2] Eventually, however, this changed, and a growing number of creationists began to claim the mantle of science.[3] Movements such as the *flood geology* movement began to emerge. Flood geology seeks to explain geological phenomena that point to an old Earth as really being evidence of the great flood for which Noah built his ark, as recounted in Genesis.[4] The goal of many of these movements was to use science to support the idea of a young Earth (this was certainly true of the flood geologists). Even those who accepted an old Earth sought to disprove the validity of evolution through natural selection.[5]

This shift represented a new shift in religious apologetics. Religious apologists, such as Cambridge University professor, Anglican clergy member, and naturalist William Paley, presumed that God was the driving force behind the natural world and focused their work on describing that world in detail. God explained the complexity they saw in living organisms and the cosmos, but there was no reason to deny that God did the creating. Paley and other – I will use the term *natural theologians* – documented the natural world, but their goal in doing so was not to prove a religious point or disprove a scientific one.[6] They had no scientific basis to doubt a divine creator at that time, just as in earlier eras, those studying the cosmos had no way of knowing that the Earth was not the center of the universe.

The natural theologians generally accepted the cosmological revolution that no longer placed the Earth at the center of the universe, and in fact they may have accepted Darwinian evolution if it had existed in their time. In the latter half of the nineteenth century, after the Darwinian revolution, many religious scientists and thinkers did not see an inherent conflict between evolutionary science and religion.[7] As we saw in Chapter 4, this remains true today for many people of faith.

The use of science to prove creation, as opposed to presuming that creation and science were compatible, became more of an issue as Darwin's work gained greater acceptance. Even then, many creationists sought to exclude evolution in the name of religion rather than claiming a scientific grounding for creation.[8] In fact, this was the sort of approach taken by the Arkansas law struck down in 1968 in the *Epperson* case.[9]

Still, by the time of the *Epperson* case, the concept of scientific creationism had grown into a full-fledged movement, with various subgroups, offshoots, and publications.[10] Much of the movement's work, especially flood geology, had gained no acceptance within mainstream scientific circles, and the movement was mostly composed of people without serious scientific training or with training in fields that had little to do with their creationist research.[11] The work of these scientific creationists was met with ridicule when it was discussed at all by mainstream scientists, not because it was religiously based but rather because it was very bad science.[12] There was virtually no original research or serious use of the scientific method.[13] There was a certain amount of preaching to the choir; that is, creation science was, and still is, wildly popular among biblically literalist Christians and largely ignored or scoffed at by most other religious thinkers and scientists. In fact, as Ronald Numbers explains, prior to the resurgence of young Earth creationism and creation science, many creationists had accepted the notion of an old Earth based on geological science and potential readings of scripture that were consistent with an old Earth.[14]

Human evolution, however, posed a challenge even to these more scientifically enlightened creationists,[15] a challenge that remains today.

The creation science movement was, in the apologist tradition, about proving the validity of faith, in this case a specific biblically literal sort of faith. Fund-raising was done primarily among like-minded religious groups. Like-minded religious individuals and groups were the target – or at least the primary audience – of creation science publications.[16] Some of the most interesting battles arose as the result of theological tension or downright religious bigotry within the movement itself. For example, a number of Seventh Day Adventists were heavily involved in flood geology, which was originated by a Seventh Day Adventist adherent, George McCready Price, but many of those most involved in promoting flood geology were from fundamentalist Protestant traditions.[17] Some were welcoming of Seventh Day Adventists, but others were not so kind.[18]

The great irony is that these issues arose in a supposedly scientific movement. If the research was solid science, why should anyone care about the religious denomination of the researcher? One issue that was relevant was the fact that most Seventh Day Adventists then, as is still true today, were supporters of religious freedom and especially the concept of separation of church and state.[19] Price did not want creationism taught in schools (he also did not want evolution taught) because he believed that would violate separationist principles, and in his view, teaching religion was not the job of government entities.[20]

Of course, unlike intelligent design (ID) advocates, neither Price nor his opponents in the burgeoning creation science movement had any reason to deny the religious roots of their approach. For Price, creationism was religiously based and therefore not the business of government-run schools.[21] For some others in the movement, creationism was religious and precisely for that reason needed to be taught in the public schools to offset the increasing trend of including evolution in biology.[22]

I point this out because many people seem to think that creation science is either a very old or a very new concept. Some presume that it has always been around, often confusing natural theology with creation science, and others believe it was a response to the *Epperson* case. Yet creation science in its modern form is a bit over 100 years old.[23] It was a reaction to the increased support for Darwinian evolution. Though some of its strategies eventually became focused on legal results, its substance and internal disputes were based on theological precommitments that run contrary to modern science. The form ID has taken, however, seems directly driven by the Supreme Court's decision in *Edwards v. Aguillard* in 1987. After all, why deny that Big D is God – when your fundamental strategic document acknowledges that Big D is God[24] – if you are not concerned about getting your approach into the schools?

I. From Natural Theology to Madison Avenue

During the period when natural theology in one form or another reigned (most of history), science was able to study the features of things in the natural world, but when it came to explaining how complex life-forms came into being, the role of God, or earlier in the history of natural theology, gods, was presumed. Even when it came to cosmological phenomena, God's role was presumed.[25] This makes a great deal of sense as science and scientific tools were not anywhere near as advanced as they are today. Explaining the features of a natural phenomenon and maybe even how it functions was, and remains, a significant achievement. Explaining the processes by which such complexity came about, whether it be nuclear fusion in stars or evolution by natural selection in organisms, was beyond the ability of science throughout most of history.

Some have asserted that the presumption of God as the creator prevented such advances. I would not hazard a guess on such an assertion other than to say that certainly the industrial revolution and post-Enlightenment thinking about our place in the universe played significant roles in spurring modern science – and aspects of post-Enlightenment thinking certainly had a role in not presuming that a supernatural force was both the origin and process by which complex (and noncomplex) natural phenomena came to be.[26] This may be something of a chicken-and-egg question, but even in the era of natural theology, it seems that great minds such as Galileo[27] and Giordano Bruno[28] saw no inherent conflict between God and science as an explanatory mechanism for natural phenomena. Unfortunately, during that time, the Church had little tolerance for such ideas,[29] but as you learned in Chapter 4, that has since changed dramatically.

The key is that there was no need to prove that God was the creator because it was simply assumed. Religious apologetics certainly existed writ large, but science and religion were not completely separate spheres during those times, so the need to prove religious truth (as already assumed) via science was unnecessary. Plato, Cicero, Aquinas, Paley, and others needed no marketing strategy to promote their ideas. By the time of Paley, those ideas were widely assumed. Paley's great contribution was his meticulous cataloging and description of natural organisms, not his use of the watchmaker analogy, an idea that existed in one form another at least as far back as Plato.[30] When scientific ideas began to challenge religious doctrine in post-Enlightenment society, burning people at the stake for their ideas (or using fear of such repercussions to scare them into silence) was no longer a valid option.

Whereas society had already come to accept the cosmological findings of Copernicus, Galileo, and others, Darwin posed a new challenge.[31] His ideas challenged directly the creation story in Genesis. Darwin considered his ideas within the scientific realm and did not assume that his ideas would be the end of religion,[32] but certainly some in the religious community viewed his ideas as a direct threat to their beliefs and values.[33] As noted earlier, some saw no conflict between his ideas and faith or biblical truth, but others – especially biblical literalists – found the idea of natural selection to be a rejection of God and the Bible.

Significantly, though, the issue remained openly focused on God. There was no attempt to hide this fact, and whatever one thinks about creation science, which was the precursor to ID, that movement was and is honest about its religious roots and assumptions.[34] This is even more true of traditional young Earth creationists, who reject the mantle of science generally when it conflicts with biblical creation.[35] It is also true of old Earth creationists generally, who accept geological science as not directly in conflict with scripture but who generally reject evolutionary biology, or at the very least evolution as applied to complex organisms, especially humans.[36] One can accept or reject the tenets of young Earth creationism, old Earth creationism, and/or creation science, but one knows what those tenets are, and those who follow these beliefs see no reason to hide those tenets. All these groups may attack evolution, but they do so from an openly religious worldview. As a legal scholar who has studied these movements, even if I do not agree with their assertions, and perhaps because those assertions have proven to be a legal liability, I find their honesty refreshing!

The ID movement has abandoned openly acknowledging its ties to natural theology or religion generally. As we have seen, this has not always been the case, and in fact, there is significant documentation of the religious core of ID, at least in documents and statements that were not intended for the public.[37] Moreover, the tenets of ID are remarkably similar to those of natural theology (for ID's positive arguments) and creation science (for ID's arguments against evolution). Whatever one thinks about the scientific or intellectual merit of ID's assertions regarding science, or the preconceptions of some of its advocates, one would have to be quite naive to deny the fact that ID advocates have learned a lot from the world of marketing, public relations, and advertising.[38] While the work of ID may never have a place in natural history museums, it should have a place in the history of great marketing strategies. ID would be quite at home on Madison Avenue.

If one believes that evolution is a threat to faith and that it stands on shaky scientific ground, and one understands that religiously grounded "scientific" theories are unlikely to win either legal or scientific battles, how might one respond? Phillip Johnson faced exactly this conundrum, and he responded by writing *Darwin on Trial*, a book that used legal-style argumentation to attack Darwinian evolution and decry the court cases that rejected religious alternatives to evolution through natural selection.[39] As we saw in Chapter 2, despite his suggestions to the contrary, Johnson takes on the role of advocate, not judge, in the book. He advocates against Darwinian evolution and attempts to poke holes in it and its acceptance by mainstream science.[40] Needless to say, when his book was reviewed by scientists, he lost the case in the court of science and Darwin won the trial, both because the scientific assertions in Johnson's book were factually inaccurate and because Johnson is a law professor, not a scientist.[41] Quite the opposite was true of the reviews of *Darwin on Trial* in creationist religious literature.[42]

As the template for a marketing strategy against evolution aimed not at scientists, but at the general public, and especially those who already question evolution for religious reasons, Johnson's book was a triumph. If I held Johnson's views of the world – which I do not because I cannot ignore or brush over the evidence supporting science or the reality of religious hermeneutics; that is, I believe a bit of skepticism is helpful in all endeavors whether natural or metaphysical – I would view his book as a template for critiquing evolution. Not surprisingly, a number of bright folks, who also happened to share similar religious precommitments with Johnson, saw it as exactly that.[43] More important, they saw it as a welcomed entry in the birth of a new movement directed at taking on that old nemesis of supernatural causation, evolution through natural selection.[44]

As these individuals began to form a movement, they needed three things. First, they needed more people, and especially people with scientific backgrounds or at the very least academic backgrounds.[45] The inclusion in this early group of Michael Behe gave them hope that this was possible. Second, they needed funding and other support. This would soon come from the Discovery Institute.[46] Third, they needed a strategy that might allow them greater success in legal cases and in the court of public, and maybe even scientific, opinion.[47] This came in the form of the wedge strategy initially developed by Johnson but that was never intended to be openly discussed in public.[48] Though this strategy belies some of ID advocates' most ardent claims that ID is not a religious movement, it is not a bad strategy from a marketing perspective if one's long-term goal is to win in the court of public opinion. Those supporting the wedge probably overestimated its likelihood of creating change in the law and in the sciences, at least in the short term (the initial timetable set forth in the Wedge Document).[49]

Yet as a long-term plan to influence as many minds, especially young minds, as possible into doubting the veracity of scientific proof and eventually to seed the scientific and legal communities with as many adherents as possible, the wedge strategy seems a reasonable one. Of course, it, too, was honest about its end goal – to defeat scientific materialism and support a Christian worldview.[50] The movement it helped spawn, however, came to understand the pitfalls of such honesty in the legal and scientific arenas.

Exit God, enter Big D. One of the central arguments of the ID movement is that the designer need not be God.[51] Moreover, another major implication of the ID approach is that it is not biblically based.[52] This clearly distances it, at least in the open, from creation science. Finally, another key aspect of ID is that the movement is primarily made up of old Earth creationists who do not generally buy into flood geology or some of the other approaches advocated by creation scientists.[53] This is, of course, not a unanimous sentiment, even in the ID movement,[54] but you will rarely find any of the main ID proponents openly advocating for a young Earth. It is true, however, that the ID movement was born when Phillip Johnson was able to broker a truce between young Earth and old Earth creationists,[55] but nonetheless, the movement is not keen on supporting young Earth ideas in public because doing so would obviously undermine its claims to the mantle of science.

By denying that the designer is God, ID advocates are able to make some interesting arguments. First, they claim that ID is not a religious concept.[56] Second, if their approach is not about God, it is easier to argue that it is an alternative scientific theory rather than the latest creationist alternative to mainstream science. Third, claims of discrimination made by ID advocates have more public and rhetorical appeal because if ID is not religiously based, excluding it from the scientific debate seems more onerous.

Yet the evidence has strongly pointed to, and still points to, ID being a religiously affected approach. This is so not just because of some of the off-the-record statements made by ID advocates acknowledging this,[57] or the fact that the Wedge Document makes this even more obvious,[58] or even because of the remarkable resemblance of ID's core beliefs to those of natural theologians such as Paley;[59] rather the simple fact that ID proclaims a supernatural designer regardless of the name given to that designer makes ID a religious approach as a matter of law and common sense.[60] From a legal and scientific perspective, this seems a rather open-and-shut case: supernatural causation equals religion.

Significantly, however, ID advocates need not accept defeat so easily. The marketing strategy is not aimed just at scientists and judges. It is aimed at the general public and even at politicians. If ID advocates can persuade these individuals, who knows what might happen in future court cases and public debates about science. After all, judges do not always apply the law in the most obvious ways in establishment clause cases, and the general public is clearly not well versed in the natural sciences.[61] This is the brilliance of the ID marketing strategy. It can adapt to legal and scientific losses while still working to persuade those with little legal or scientific background of ID's place in the sciences, or at the very least that ID and ID advocates are the victims of pernicious discrimination in scientific and legal fora.

An exceptionally good example of this is the ID movement's response to the decision in *Kitzmiller*. As you learned in Chapter 3, *Kitzmiller* was both a legal and scientific defeat for ID. The court, after hearing testimony from leading thinkers on both sides and reviewing mountains of evidence, held that ID is not science and that it is religiously grounded.[62] Initially, ID advocates had asked Judge Jones to address whether ID is science.[63] They seemed to assume that because Judge Jones was appointed by President George W. Bush, himself a supporter of ID, the issue would be determined in their favor.[64] Of course, anyone who looks clearly at the testimony and other evidence in the case can see how any reasonable judge would conclude that ID is not science. A judge would have to ignore the clear weight of the evidence regarding ID, regardless of the behavior of the school board, to find otherwise.

As more facts about the behavior of the Dover School Board came to light, the Discovery Institute and some of its associates backed away from the case. This was a wise decision. The choice was simple: be tied to the Dover School Board and lose horribly or distance yourself and fight again later. Still, Judge Jones did not base his decision about the nature of ID on the behavior of the school board but rather on all the evidence discussed in Chapter 3 and the *Kitzmiller* opinion itself.[65]

The response to the *Kitzmiller* decision by ID advocates has been fascinating. They have attacked it from numerous angles. First, they have engaged in an organized campaign to discredit the decision. The Discovery Institute has led the charge by attacking Judge Jones, the opinion in the case, and the Dover School Board.[66] Discovery Institute fellows and staff have called Judge Jones a "judicial activist,"[67] attacked his credibility,[68] attempted to undermine the breadth and depth of the evidence in the case,[69] mocked the legal decision,[70] and attempted to distance ID from the behavior of the school board.[71]

This is remarkable given the movement's predecision treatment of Judge Jones and the defendant's call for the court to address the question of whether ID is science.[72] It seems that Judge Jones, who was described by ID advocates before the decision was issued as being a good ole boy Bush appointee, a religious man, and a judicial conservative,[73] somehow became a judicial activist and friend of the Left overnight.[74] Some allies of the ID movement even called him antireligious.[75]

Yet Judge Jones explained the basis for his decision in detail in the opinion and in later interviews.[76] It was the evidence that persuaded him to rule as he did.[77] He explained in subsequent interviews that he had started out believing that ID advocates would be able to make a powerful case.[78] Of course, that never happened, and the evidence in the case supporting ID as a religiously based approach as well as the evidence demonstrating that ID is not science was overwhelming.[79] Also, the allegations that Judge Jones is antireligious – which were not made directly by Discovery Institute representatives – have absolutely no basis in fact. Judge Jones was never viewed as antireligious by ID advocates until they lost the case.[80]

From a marketing perspective, *Kitzmiller* was a devastating blow to ID, not just because of the substance of the decision but also because it laid bare to public scrutiny some of the movement's dirty secrets such as the evolution of the book *Of Pandas and People*[81] and the Wedge Document. Lest we forget, however, ID advocates are resilient and brilliant marketers. Attacking the decision and Judge Jones in the manner they did strains legal credibility (I am here referring to the attacks marketed to the general public; we will address the legal analysis set forth by ID movement lawyers in response to the decision later). Yet it is a sound marketing strategy. It is a classic case of attacking the messenger. In this case, that tactic is good marketing because the message in the case, for those who have actually read it and looked at the evidence, is damning to the ID movement's assertions that it is scientific and not religious.

Significantly, such attacks are also an excellent way to gather the troops and raise money. Incredulity at the supposedly activist decision of a Godless federal judge banning Big D from the public schools of Dover is an excellent strategy to raise money. Think about it. Whether the decision was actually activist or not, whether Judge Jones is really antireligious or an activist judge or not, the allegations when heard by those who already agree with the ID movement's endgame and apologist approach will raise ire, support, and funding.[82] It also plays beautifully into the victim status claims made by ID advocates that were already in play before the *Kitzmiller* case was even filed. The scientific, religious, educational, and philosophy experts, along with the plaintiffs, their lawyers, and the judge, now become a cabal imposing unfair and discriminatory treatment on ID.

Moreover, lawyers affiliated with the ID movement have attacked the decision, and Judge Jones, in law review articles.[83] Most of these articles attack the decision for reaching the question of whether ID is science and/or religion.[84] Relatedly, the articles question the court's analysis regarding the nature of science and the nature of ID.[85] They also question the court's reasoning under the establishment clause.[86]

Most of the legal arguments raised in these articles have been discussed in earlier chapters. However, accusations in the articles and press releases by ID advocates that Judge Jones's decision was activist must be addressed. What is judicial activism, and is *Kitzmiller* an activist decision?

The definition of *judicial activism* has been debated for decades. Most legal scholars agree that overturning laws passed by the democratically elected branches of government without good reason – such as a constitutional violation – is at the core of judicial activism.[87] Others add that following precedent, known as the principle of *stare decisis*, is central to avoiding judicial activism.[88] It should come as no surprise that many accuse courts of judicial activism when they do not like the results and/or reasoning in a court decision. There are situations in which both sides of a legal debate have accused courts of activism even though the courts themselves cannot agree on a result; one side views the courts that held against its position as being activist, and the other side views the courts that held against it as being activist. There are also frequently political allegations made about activism such as liberal judges are activist or conservative judges are activist. As I tell my students, judicial activism is not the province of political liberals or political conservatives; rather it is the province of activists regardless of their political leanings.

Whatever definition one uses, however, it is quickly apparent that the *Kitzmiller* decision is not an example of judicial activism. First, the decision is clearly based on the evidence presented at trial. Second, the analysis of the legal issues under the establishment clause is consistent with Supreme Court precedent and the evidence presented in the case. Third, the defendants asked the court to rule on whether ID is science, and in fact, that was the focus of much of the evidence presented by both parties. The court did not simply assume an answer to a question that was not properly before it or on which evidence was not presented.

The court did, however, overturn a policy passed by a democratically elected body. Under the circumstances, however, where the evidence supported a finding that the policy violated the U.S. Constitution, activism is not a well-supported allegation. Moreover, the behavior of school board members, which included lying on the stand, strongly supported the conclusion that the motives and actions of the elected officials in this case were not entitled to the usual deference and did not necessarily reflect the will of those who elected them. This was clearly demonstrated when the board members who supported the ID policy were voted out en masse in the next school board election.[89]

After the legal defeat ID suffered in *Kitzmiller*, the Discovery Institute realized that a public triumph of some sort would be helpful, and they wisely turned their attention to the political process and the state of Kansas. That state was embroiled in a battle (for the second time) over the state school board's attempt to include ID and the notion of teach the controversy in the state science curriculum.[90] The board was stacked in favor of ID, although dissenting board members sometimes did speak out.[91]

Several years earlier, the Kansas State Board of Education had promoted a similar idea. This led to a significant response by the scientific and academic communities.[92] A policy favorable to ID was passed, but the board that passed it was quickly voted out of office in the next election cycle, and the rule was revoked.[93] Interestingly, the Kansas State Board of Education had a large contingent of creationists each time these policies came forward.

If ID is not a religious approach, why is it generally brought forward by creationists in the political process? Dover and Kansas are two examples.[94] Moreover, why is it generally biblically literalist religious entities that speak out in support of ID in such circumstances rather than scientific entities or religious entities that are not biblically literalist?

This time, however, the scientific community in Kansas and elsewhere decided to boycott the hearings. There were two primary reasons for this. The first was political. It seemed clear that science was not likely to get a fair hearing before the board committee in charge of the hearings.[95] The second reason demonstrates that the scientific community was getting more savvy in addressing the marketing strategy favored by ID advocates. Every time one of these hearings takes place, the scientists reasoned, we are pitted against ID as though we are scientific alternatives or even equals.

This is the brilliance to the ID movement's teach the controversy approach. Anytime ID can share the stage with mainstream science, ID is able to gain a bit more cache.[96] To do so at the state board of education level, which was going to be covered by the national and international media, would give ID a level of credibility the scientists asserted it had not earned.[97]

School board members decried the decision of the scientific community to boycott the hearings, but scientists did provide commentary to the media (which board members also decried).[98] The hearings, which were recorded, are remarkable for the fact that board members praised the ID advocate who testified, while making incredibly uneducated statements about science and the educational process.[99] Also, the creationist leanings of the board members were readily on display, which did cause some of the ID advocates a bit of embarrassment as they tried to distance themselves from traditional creationism.[100]

The end result of the hearings was the passage of a new policy that favored ID.[101] The scientific and academic communities, as well as many parents who were concerned about their children's ability to learn science and get into and succeed in postsecondary education, complained.[102] As had happened before, the next state school board election shifted the balance of power on the board, with some of the creationist board members being voted out of office.[103] Therefore the policy never took effect.[104]

The ID movement has faced legal and political defeats, but its resiliency and marketing strategy should not be underestimated. The ID movement has been publicly supported by a sitting president, George W. Bush. Additionally, a well-known television personality, Ben Stein, has supported ID, going so far as to make the movie *Expelled: No Intelligence Allowed*, discussed elsewhere in this book.[105] Moreover, the ID movement has enjoyed support from other television venues, especially FOX News.[106]

The ID movement also enjoys significant support in some religious communities, which is, as others have suggested, something of an indicator of its core and message. Funding has not been lacking,[107] and the Discovery Institute seems poised to continue its assault on evolution and promotion of ID for years to come. New strategies are evolving, and older strategies are being revitalized. Those who believe that *Kitzmiller* and Kansas portend the end of the ID movement may not have noticed that there are still creation scientists out there, even after *Edwards*[108] – although as a legal matter, they have become an afterthought.

ID advocates may or may not realize that their marketing strategy is likely to be unsuccessful in persuading those who understand science and the law that ID has a place in science or a place in government-supported science classrooms, at least in the near future, but the marketing strategy is not just a short-term one.[109] Whatever else one might say about the ID movement's substance – and this book has said a lot – one thing has clearly been intelligently designed, and that is the ID movement's strategic marketing approach.

II. Next Move?

Looking toward the future, the ID movement is promoting several approaches. The primary approach in the legal and political context is retooling and strongly marketing the teach the controversy argument. Another aspect of the teach the controversy approach is supporting teachers who want to teach ID but are told not to do so by school officials. ID advocates also continue to tout the scientific arguments for ID – thus the rise of the Biologic Institute, which ID advocates point to as evidence of their continued scientific agenda.[110]

These three approaches provide a glimpse of the direction the ID movement is going in the second decade of this century. Of course, all three approaches have significant marketing power. The teach the controversy approach as well as support for teachers who want to teach ID and are denied the ability to do so play beautifully with those who already support ID and may create a sense of righteous indignation in others who hear only the ID movement's side of the story. The Biologic Institute and its research agenda have the potential to be immensely popular among those who already support ID, but even more important from the perspective of ID advocates, its scientific pitch has the potential to be persuasive to those who know little about current science and the scientific method, that is, almost everyone. We will turn to the third approach first.

A. The sound of science

As Ronald Numbers has explained, for many years, creation scientists struggled to find people with serious scientific credentials to back the movement.[111] They were able to get a few people with credentials, but that did not save the science part of creation science because there was no success in proving anything convincingly to the scientific community. Perhaps the most credentialed member of the movement, Kurt Wise, who studied under famed biologist Stephen Jay Gould, understood the tension between his creationist beliefs and the mainstream scientific world.[112] By all accounts, Wise is an immensely principled human being, and he seemed to know that good science – by the standards of mainstream science – did not support creationism, even creation science.[113] By taking a teaching post at Bryan College, a Christian institution named after William Jennings Bryan, where Wise also directed the Origins Research and Resource Center, despite his Harvard PhD and work under Gould, it appears he decided that given the choice between abandoning his faith in creationism or his position within mainstream science, only the latter was a viable option.[114] Whether one agrees or disagrees with Wise's decision or beliefs, it is hard not to respect the courage of his convictions.

The beginnings of the ID movement also saw an attempt to bring in highly credentialed individuals in the sciences, and like creation science, the ID movement had little success in this endeavor, but the early ID movement did include a few people, such as Michael Behe, with strong mainstream scientific credentials. This revealed another problem, namely, that even these people were unable to get ID material published in peer-reviewed journals.[115] They were, however, able to get non-ID-focused research published.[116] They argued that this work supported ID, but doing so required a deductive leap that was scientifically unsupported based on the publishable research alone.[117] This was revealed to devastating effect in the *Kitzmiller* trial.[118]

In 2005, the Biologic Institute was founded. It is heavily supported by the Discovery Institute.[119] It has done what predecessors have failed to do. The Biologic Institute has been able to bring together perhaps the best credentialed group of individuals the ID movement has ever been able to collect. Moreover, the institute has added a new twist. In addition to focusing on biological systems, the institute is focusing on the parallels (as asserted by the institute) between complex technology designed by humans and the complexity found in nature.[120]

From a marketing perspective, all this is priceless. The ID movement should get a lot of mileage from this new institute. As will be seen, however, from a scientific perspective, the Biologic Institute's approach suffers many of the same flaws suffered by complexity by design and irreducible complexity.

The Biologic Institute was founded to support design scientifically. In the Institute's own words,

> Biologic Institute is a non-profit research organization founded in 2005 for the purpose of developing and testing the scientific case for intelligent design in biology and exploring its scientific implications.[121]

The institute's home page lays out its approach in exceptionally well-developed – in the sense that any public relations expert would be proud (my characterization, not the institute's) – language:

> Biologic Institute brings together scientists with diverse expertise, unified by the realization that a revolution in biology – with far reaching implications – is well under way. Like many revolutionary ideas, this one is powerful in its simplicity: The more we learn about the organization of life, the more clearly it reveals design.
>
> When you realize that living cells store, transmit, and process information, the similarities with human technology are unavoidable. But when you get a glimpse of the remarkable sophistication of the cellular processes – and the almost unbelievably small scale of the molecular systems performing them – you begin to realize that humans are novices when it comes to complex design.
>
> If you're like us, you also begin to think about the exciting possibility of bringing these two worlds together: the world of human designs and the world of living designs. Biology

> is already informing technology, and we think the reverse will prove true as well.[122]

Wow, sign me up! But wait, now I am starting to think . . . hold on a second . . . there are a lot of assumptions in the preceding statements! In fact, now that I think about it, are the preceding assertions not just Paley's watchmaker on steroids?[123] Is this not just a rehash, with a technological twist, of the sorts of arguments the *Edwards* and *Kitzmiller* courts found to be religiously based? The Biologic Institute addresses this, stating that its approach is scientific and implying that it is not religious, as it sets forth its view on the ID controversy:

> Our take on the ID controversy
>
> "Intelligent design" refers to the claim that many aspects of the universe and of living things are best explained by an intelligent cause.
>
> The significance of this claim, if true, is not controversial. In the words of eminent Harvard biologist E. O. Wilson, "Any researcher who can prove the existence of intelligent design within the accepted framework of science will make history and achieve eternal fame." According to him, even the Nobel Prize "would fall short as proper recognition."
>
> Notice that Wilson here concedes that proving the design hypothesis is not categorically unscientific – if it were his statement would be nonsense. Furthermore, most careful thinkers concede the reasonableness of inferring that life *was* designed.

Geneticist Graham Bell, for example, writes the following in the introduction to his excellent technical monograph on natural selection:

> A light bulb or a lathe are preconfigured in the mind, and constructed according to a plan. It is entirely reasonable to assume that beetles and daisies must be constructed after the same fashion, especially because they are much more complicated than anything that human ingenuity has so far managed to devise.

Notice the absence of any mention of religion here. Bell seems to be saying that inferring intelligent design is just plain reasonable. Like Wilson, he prefers Darwin's explanation....

So, in view of the unmistakable importance, the scientific legitimacy, and rationality of this design idea, you'd think every scientist would be asking the obvious scientific question – Is it true?

Many are, of course, and in this they share our judgment that the best way to settle scientific disputes is by doing good science. Regrettably, though, others seem bothered by the suggestion that this dispute hasn't already been settled in Darwin's favor. These people are happy to let science settle the matter – provided we all submit to their pronouncement that the case has been closed.

But science doesn't work that way. It never has.

Science is the open-ended, open-minded pursuit of the best data and the best interpretations of those data. That's

> the public enterprise that the public funds for the common good. So, instead of lamenting the fact that Darwin's idea has never achieved a convincing win on the public stage, perhaps his sympathizers would do well to ask whether they really have a case. If convincing science is very convincing (and it is), then an unconvinced public should perhaps tell us something.
>
> The scientists at Biologic Institute are convinced that careful science on the nature of life is poised not only to answer the design question, but also to usher in an exciting new era of scientific discovery.[124]

This statement is powerful and compelling if one knows little about science and even less about the works and people quoted in the statement. Let us dissect the statement to look at what it really says and what it leaves out. This statement has all the elements of the ID marketing plan: denial of a religious motive, the idea that conclusory statements and deductive reasoning of the sort in the statement have scientific value, allegations of closed mindedness and exclusion by the scientific establishment, the assertion of a scientific controversy, the notion that public sentiment against a scientific theory says anything about the validity of the scientific theory (remember that public sentiment long favored a flat Earth and the Earth as the center of the universe), assertions about the inescapable logic of design given the complexity in living things and the universe, and the use of quotes by mainstream scientists – taken out of context – to support the notion of design, in this case, E. O. Wilson and Graham Bell.

Let us parse the preceding text. The first sentence is certainly accurate enough. The next sentence involves a sleight of hand that sets up everything that follows: "The significance of this claim, if true, is not controversial." Notice the words "if true." The significance of any claim "if true" would be greater than if the claim were untrue. Moreover, while the significance of the design claim would not be controversial "if true," the same could be said about the Earth as the center of the universe if it were true. This statement presumes a lot, namely, that there is some reason to believe scientifically that ID is true, but as we have seen, this is not the case. Still, let us give the Biologic Institute the benefit of the doubt on this sentence and read the statement to mean that they plan to prove scientifically that it could be true. Significantly, the rest of the statement betrays this position.

Next comes the quote from E. O. Wilson, who, as the sentence suggests, is indeed an eminent Harvard biologist. He also completely rejects the notion of ID and in fact is not terribly sympathetic to the notion of theistic evolution.[125] In fact, Wilson, like Dawkins, seems to believe that faith and evolution through natural selection are inherently in tension with each other.[126] There is no question that Wilson views intelligent design as being on the faith side of that equation, not the scientific one.[127] Recall the quotation (partially paraphrased) from Wilson on the institute's Web site:

> "Any researcher who can prove the existence of intelligent design within the accepted framework of science will make history and achieve eternal fame." According to him, even the Nobel Prize "would fall short as proper recognition."

Now let us view that quotation in the context in which it was written. It hardly supports the assertion made on the Web site that "Wilson here concedes that proving the design hypothesis is not categorically unscientific – if it were his statement would be nonsense."

The Wilson quotation in context is as follows:

> Flipping the scientific argument upside down, the intelligent designers join the strict creationists (who insist that evolution never occurred) by arguing that scientists resist the supernatural theory because it is counter to their own personal secular beliefs. This may have a kernel of truth; everybody suffers from some amount of bias. But in this case bias is easily overcome. The critics forget how the reward system in science works. Any researcher who can prove the existence of intelligent design within the accepted framework of science will make history and achieve eternal fame. They will prove at last that science and religious dogma are compatible. Even a combined Nobel Prize and Templeton Prize (the latter designed to encourage the search for just such harmony) would fall short of proper

> recognition. Every scientist would like to accomplish such an epoch-making advance. But no one has ever come close, because unfortunately there is no evidence, no theory and no criteria for proof that even marginally might pass for science.[128]

Is Wilson suggesting here that the "design hypothesis" is scientific? Does this quotation suggest that most "careful thinkers concede the reasonableness of inferring" intelligent design? It would seem, based on the full paragraph from which this quotation was taken, that Wilson rejects both these claims (his work as a whole clearly rejects these claims). Two possibilities exist here. First, Wilson is not a careful thinker because he rejects any implication of design, in which case, why should we pay any attention to the quotation the institute uses to support its enterprise, and why, if he is not a careful thinker, would the institute tout his credentials and the quotation itself? Perhaps he is a careful thinker but somehow missed the design path that supposedly should be obvious to most careful thinkers? Who are these careful thinkers anyway?

We know that virtually all scientists reject the reasonableness of inferring design. The same is true of many philosophers of science, religious thinkers, and religious denominations. Are none of these people or institutions careful thinkers? They would seem to be the thinkers most relevant to the underlying question. Perhaps to be a careful thinker, one must already agree with the design hypothesis?

The second possibility is that Wilson *is* a careful thinker and rejects the assertion attributed to him, but the statement on the Web page is basically just a marketing ploy designed to make it seem like Wilson supports the institute's assertion. This seems far more plausible and is inescapable when we view the rest of the statement. Wilson certainly does not view the design hypothesis as scientific. He says so specifically in the last line of the preceding quoted statement and in a number of other writings.[129]

Wilson's statement is not "nonsense" as a result of this; rather, and quite ironically, it seems the whole point of the quoted paragraph (and the section of Wilson's book from which it was excised by the institute) is to demonstrate that ID is not scientific. It is true that Wilson does not say that ID is "categorically" unscientific, but what does that mean? It means he, like any scientist, is unwilling to say that there is no sort of proof that would make it scientific.[130] Wilson is saying, however, that such proof would have to meet the requirements of science and that no such proof has ever been demonstrated or is likely ever to be demonstrated, and perhaps most important, that the approach of the ID movement could never provide such proof because the movement's approach (which presumes design and goes from there) could never meet the requirements of current science.[131] Some ID advocates have acknowledged that for ID to be science, the definition of science would need to change.[132]

It should not be overlooked, however, what a brilliant marketing strategy this is: make unsubstantiated statements that have a great deal of commonsense appeal; next, use the name and words of a prominent Harvard scientist to support your point, albeit by taking that scientist's words out of context and making them seem to ratify propositions that are the complete opposite of those actually stated when the quotation is read in context; and, finally, suggest that anyone who does not agree with your ideas is not a careful thinker like this Harvard scientist and whoever else counts as a careful thinker. The great irony is that the Harvard scientist is a careful thinker, at least when it comes to science, and he disagrees that careful thinkers find inferring design to be reasonable. Again, though, the facts are quite irrelevant to the purpose of such statements.

This is solid market puffery, or even classic public relations spin/fabrication. If one does not take the time to check the context of the quotes, the veracity of the assertions on the Web page, or the list (or lack thereof) of supposed careful thinkers or to parse the language in the statements to determine if there is any difference between calling something unscientific versus "categorically unscientific," this language can support a positive view of the design hypothesis. This is one of the reasons I quote the institute's statement and parse it here. It gives us great insight regarding the use of language and issue framing by ID advocates to support their case in the court of public opinion.

In addition to Wilson, the page quotes McGill University geneticist Graham Bell. Again the quote is taken completely out of context, and the suggestion is made that it stands for a proposition that anyone familiar with the broader context of the quotation or the work of Bell generally would find laughable.[133] Then again, most people reading the Web page would not be familiar with the book from which the quotation is taken or Bell's work generally. The page quotes Bell as follows:

> A light bulb or a lathe are preconfigured in the mind, and constructed according to a plan. It is entirely reasonable to assume that beetles and daisies must be constructed after the same fashion, especially because they are much more complicated than anything that human ingenuity has so far managed to devise.

The quote is taken from Bell's book *Selection: The Mechanism of Evolution*.[134] It is taken out of context. In fact, Bell not only rejects design arguments in the next paragraph, but the book itself is written to explain the importance of natural selection in biological processes. The quote in its full context reads as follows:

> Few of us would be able to design a light bulb or a lathe, still fewer the computer and its attendant software with which this sentence is being written. But we all have a clear idea of what is meant by "design," and we readily, too readily, transfer this notion to the natural world. A light bulb or a lathe are preconfigured in the mind, and constructed according to a plan. It is entirely reasonable to assume that beetles and daisies must be constructed after the same fashion, especially because they are much more

complicated than anything that human ingenuity has so far managed to devise. There is, however, a second route to complex organization, through the selection of random variants that propagate nearly exact copies of themselves. It is of very little consequence in our daily lives, because it is so much more laborious and expensive than deliberate design. However, it is another way of constructing things. Indeed, so far as I know, it is the only other way of constructing things that we have ever been able to imagine. Selection is not merely a theory antagonistic to some other possibility, in the same way that oxygen and phologiston or open and closed circulatory systems are antagonistic theories; it is one of the only two categories of theory competent to explain the living world. It is as important as all the other scientific theories combined, because it offers a different *category* of explanation.

Of course, it might be wrong, and it could be relegated to the footnotes of historically-minded textbooks of biology. There are, indeed, two influential schools committed to the footnote position. One consists of religious people who with different degrees of stridency assert on their own authority that the world was designed and created by an intelligence superior to ours. The other consists of a mixture of different kinds of people, mainly scientific, who think that life embodies a principle of inherent self-organization, with selection playing a minor role, if any. *These two schools are mistaken, root and branch. Living complexity cannot be explained except through selection, and does not require any other category of explanation whatsoever.* This is the

> position that this book sets out to assert, to elaborate, and to exemplify.[135] (Emphasis added in last paragraph)

Lest there be any doubt, Bell is not suggesting that ID is a plausible scientific theory. He would apparently place ID in the first school he mentions, that is, the religious school.[136] Anyone who would suggest otherwise might want to take the time to actually read the book. If there is any doubt that Bell might be in some way suggesting that ID is an alternative scientific view, that view is quelled in the remainder of the introduction and throughout the book, where he discusses various scientific arguments related to selection, and there is nary a mention of ID.[137]

After the inferences made based on Bell's quote, the excerpt quotes Richard Dawkins's statement that people may become religious because they perceive design in complex natural phenomena. I cut this element from the preceding quotation because having discussed Dawkins in detail in Chapter 5, the obvious ridiculousness of trying to use this logic to support design makes discussion of this in any detail unnecessary here. Suffice it to say that Dawkins's statement is part of his broader discussion of why these perceptions are completely at odds with the scientific facts as he believes them to be and are akin to believing that a phenomenon that cannot be explained based on the extant state of knowledge is the result of magic.[138]

In fact, quoting Dawkins here may be one of the few marketing errors made in an otherwise stellar display of marketing prowess. The reason for this is that a decent number of people have read, or are familiar with, Dawkins. For these people, realizing that the implications attributed to Dawkins are bogus may cause them to question the other statements on the page and do their homework, which would expose the assumptions and misstatements on the page generally. Certainly some people have read Wilson and Bell, but most folks who have done so are likely to have a preexisting interest in evolution and are unlikely to take anything the institute has to say seriously. Dawkins is a different story because more than the others, his work is aimed at a general audience, and even if many in that audience disagree with him, his readers (and those who have seen him interviewed on television) are aware of what he thinks, and they could easily tell that the institute implies that he says something he does not say.

Still, this is a minor marketing faux pas, because in the end, most people who read the Web page will not have read Dawkins or will not think it through as carefully as some other people might. Perhaps the marketing grade falls from an A+ to an A. For those who admire the marketing savvy demonstrated by the institute's Web site, there is no need to worry, because as we will see, the marketing A+ is quickly earned back!

After the mention of Dawkins comes the following: "So, in view of the unmistakable importance, the scientific legitimacy, and rationality of this design idea, you'd think every scientist would be asking the obvious scientific question – Is it true?" Wow, this is *Advertising Age* worthy![139] Having worked in the advertising world for a time before attending law school, I will use an advertising term here: this is *darned good copy*! Here *copy* means words, and specifically in this case, words well thought out to achieve a marketing goal.

What is unmistakably important, scientifically legitimate, and rational about the "design idea"? Remember that all these statements are apparently built on the preceding material, which, as we have discussed, demonstrates no such thing unless one takes all the assumptions for granted and ignores the misquoting of Wilson and Bell and misuse of Dawkins. Important to whom? Obviously it is important to the designers, but design is being pitched generally as scientifically "legitimate" and rational (apparently in the sense of inherently rational). Yet we know the very people they cite do not view it as such, nor do the vast majority of scientists and philosophers or a good number of theologians.[140] Of course, one is only likely to raise such concerns if one carefully reflects on the statement rather than taking it at face value.

It is the next passage, however, that binds what precedes it together: "you'd think every scientist would be asking the obvious scientific question – Is it true?" This, of course, assumes that design is indeed "unmistakably important," scientifically legitimate, and rational. If these things are not true, why would any mainstream scientist care, and more important, why would the design question be obvious? Yet in this last passage, the use of *scientist* and *scientific question* build in the expectation that design is science. The passage as a whole also raises suspicion about why scientists do not actually explore this "obvious" question. Of course, the answer to that question is provided throughout this book. The question is not a scientific one, at least not a persuasive one, and it is not "unmistakably important" to scientists because there is no convincing scientific evidence to support it.[141]

One could take almost any assertion and plug it into this sort of deductive formula. Take idea A, characterize it in your preferred way, and back that characterization up with quotations from wise people. Of course, imply that those quotations support your point. Never mind if that is completely inaccurate, because most folks will not check anyway. Next, deduce for your readers that your preferred way of viewing issue A is nearly inescapable, or at the very least worth exploring, and then imply that there is something amiss if it is not explored. Also, make sure to include idea A in the category (in the case of design science) in which you want it to be included. As a marketing plan, this has a great deal of potential. As a scientific argument, it is useless, or worse, disingenuous.

Next, the text suggests that "many" people are exploring the design question. It is good that they do not say many scientists because this would be immensely easy to disprove. By using the word *people*, the implication is scientists or researchers, but the problem of backing that up is avoided. Depending on how you define *many*, certainly one could say that many people work in the Discovery Institute, the Biologic Institute, and other groups that support ID. Under this interpretation, the word *many* would seem to mean more than a few. If you try to say "many scientists," however, this creates the problem that actually very few scientists care one way or the other about design (other than believing that it may be dumbing down people's understanding of science).[142] Even if you take all the people with scientific credentials working for the various ID entities, the numbers are paltry when compared with the number of scientists generally. Rather than "many," it would be accurate to write "a few." The idea of the "many," as it might be called, is also backed by the words "share our judgment." This suggests that the "many" are somehow outside of the ID research organizations. In this sense, it clearly could not be an accurate reference to scientists.

The many who share the institute's judgment apparently believe "that the best way to settle scientific disputes is by doing good science." Now this could apply to many. All scientists would agree with the latter part of this statement, but it is connected to the first part, which refers to the supposedly scientific design question. Ironically, most scientists reject design, as we saw in earlier chapters, precisely because it is not good science in their view. Again, though, the eminently reasonable assertion about doing good science is used to support the statement connecting such science to design as a key scientific question. Unless you are aware of this and stop to think about it, this language is highly compelling.

The next section drives home both the idea of exclusion – that ID is being unfairly excluded by mainstream science – and the argument that there is a scientific controversy regarding evolution and design. The section states that "others seem bothered" by the designer assertion that the "dispute hasn't already been settled in Darwin's favor." It does not take much imagination to determine who "others" are in this context. Others are scientists, primarily biologists, who reject that the design hypothesis could be a valid subject for scientific inquiry. This group includes most mainstream biologists. Assuming that they are "bothered" by the designer assertion is interesting. Most mainstream biologists do not pay much attention to ID and are probably not bothered by what they consider to be bunk, except to the extent that they are concerned it may be lowering scientific knowledge generally.[143] Others, such as Dawkins and Wilson, clearly are bothered by the assertion that (1) there is a dispute about evolution versus design and that (2) the designers suggest that dispute has not been settled in favor of evolution.[144]

This, however, is the power of the statement. As we saw in earlier chapters, there really is no dispute regarding evolution versus design. Vast quantities of evidence and data have been gathered using the scientific method supporting evolution, and no such evidence or data have been gathered for design (unless one allows the definition of science to change so that supernatural forces can be applied wherever an immediate explanation is unavailable).[145] Still, coming to this conclusion requires knowledge of extant scientific research and careful reflection on the substance of the statement in question.

Portraying the dispute as real – as ID advocates have done all along – creates a catch-22 for scientists. If scientists respond to the alleged dispute to show that there really is no dispute, ID advocates can use the response to argue that there is indeed a dispute, and better yet, ID advocates can state that scientists are engaging with them, which provides a sort of rhetorical credibility. If scientists do not respond to the alleged dispute, ID advocates can suggest that there is some sort of unfair exclusion or conspiracy taking place. This strategy is win-win for ID. It is meaningless from the perspective of good science, but it is priceless for those wishing to market ID.

Finally, notice that this dispute – which, as we saw in earlier chapters, has been settled in Darwin's favor based on the currently available scientific evidence[146] – clearly remains unresolved in the institute's statement. The bold assertion that the dispute is real and palpable is a page straight from the teach the controversy argument, yet in the same breath, the assertion supports the notion that mainstream scientists are unwilling to consider other "scientific theories." This, of course, assumes that ID is a scientific theory and that scientists would be unwilling to accept an alternative to evolution that actually was supported by good scientific evidence. This is followed by the indignant statement, "But science doesn't work that way. It never has." This all depends on whether one accepts the preceding statements in the excerpt. It presumes that ID is being excluded because it does not gel with the preexisting Darwinist cabal. The statement is substantively void, however, if the reason for rejecting ID is that it is bad science.

The next paragraph in the excerpt is the pièce de résistance. First, it states what science is in very broad terms with which it would be hard to argue but which give little in the way of specifics as to how scientific data are pursued and what sorts of "interpretations of those data" are considered scientific and why. Next, there is a subtle but absolutely important shift. The statement asserts that science is a "public enterprise" that "the public funds for the common good."

There is a certain amount of truth to this. A great deal of public funding goes toward scientific research, and the end goal is in part to serve the public good, but scientists are not charged with skewing experiments or drawing unsupportable conclusions just because the alternative may not fit what some consider the "public good." To what "public good" is the statement referring? Who gets to decide? The way it has generally worked is that the scientists do science and this ultimately leads to useful discoveries in medicine, space travel, technology, and understanding the world around us, which all serve the public good and public knowledge over time.

The next statement drives home what the writer seems to have intended by the first two statements in the paragraph: "So, instead of lamenting the fact that Darwin's idea has never achieved a convincing win on the public stage, perhaps his sympathizers would do well to ask whether they really have a case. If convincing science is very convincing (and it is), then an unconvinced public should perhaps tell us something." This is mind-bogglingly misleading if you think about what it says. First, it asserts that scientists are "lamenting the fact" that Darwinism has not "achieved a convincing win on the public stage." First of all, scientists do not generally spend their time lamenting this supposed fact. Second of all, it is not a fact. In the United States, Darwinism has not gotten a majority of the public's support, but in other industrialized nations around the world, it has garnered more. Moreover, Darwinism has certainly "achieved a convincing win" on the scientific stage, which would be the one we would expect scientists to care most about.

Think about what this statement does. It makes the public stage the relevant stage to consider what is good science. If this were really the case, then we should expect that science would still assert that the Earth is the center of the universe, that bleeding people cures diseases, and that our fates are heavily based on astrological signs. Science does no such thing. Why? Because scientists do science, and there has long been a disjunction between current public perception and scientific knowledge. Eventually the public perception catches up, but not necessarily at a quick pace.

Why would any scientists suggest that the public stage is at all relevant in determining what is good science? If you think about it, the whole idea seems inane, but if you do not think it through, or do not even notice that the term *public stage* is not a reference to the scientific stage, it is a compelling argument. The shift from the forum of science to the forum of public opinion in the statement is seamless.

The final point in this paragraph suggests that good science is "very convincing," and therefore the fact that a majority of the public does not support evolution tells "us something." It does: it tells us that despite the vast amount of evidence out there, the majority of the public in the United States does not accept evolution. It tells us nothing about the scientific validity of evolution. Many people in the American public believe in a young Earth, but does this mean that a roughly 6,000-year-old Earth is a valid scientific idea? Would ID advocates, most of whom are not young Earth supporters, be willing to say this because much of the public accepts it? Is the public now the judge of good science? If that had been the case for most of the last two centuries, we may have never developed much of the technology, medical advances, and so on, that we developed because certainly most of the public during the scientific revolution did not support many of the new ideas, sometimes because they conflicted with scripture and sometimes because the general public simply did not understand the ideas. No one should expect the general public to be experts in every scientific field, but apparently, the folks at the Biologic Institute do.

One more note is in order here. One might wonder aloud what role the creation scientists and ID advocates have had in affecting the public's view of evolution. There is no clear answer, but certainly both these movements have openly tried to persuade the public that evolution is bunk and that it is "just a theory" in the sense that scientists would say "it is just a hypothesis."

The final paragraph is the endgame sales pitch. It points out that the "scientists" at the Biologic Institute "are convinced that careful science . . . is poised" to address and answer the design question. This is a great conclusion for driving home the idea that the institute comprises a group of scientists engaged not just in science but in "careful science" and that its work is important. It also reinforces the earlier implications that the design question is a scientific one.

Why parse this statement in such a detailed fashion? The statement is a wonderful example of the ID marketing plan in action. It uses a number of tried and true apologist techniques but combines them with some new ideas and packages them in a crisp, well-designed, and well-written package.

It is striking that there is no description or discussion of real science in an excerpt openly proclaiming a focus on science. *Science*, *good science*, and *careful science* are essentially buzzwords designed to characterize science in a manner that suggests design belongs in the scientific realm and that the exclusion of design by mainstream science somehow violates the scientific ideal. As noted earlier, this only makes sense if ID advocates have succeeded in gathering adequate scientific proof to make science take notice. The use of these terms begs the very question at issue: is ID science? One could substitute UFOlogy or crystal power for design and make the same sorts of statements.

Moreover, the claims that evolution through natural selection has not been proven (to the public) echo the old arguments made by creation science and other ID advocates. This is a key element in the ID movement's argument. If evolution through natural selection were proven to be real, there would be little room for design, and worse still from the ID perspective, scientific materialism and methodological naturalism will have prevailed scientifically and would eventually come to be accepted by the public generally. Of course, evolution through natural selection has been proven to be a real phenomenon, and though public understanding lags behind scientific understanding, the public may eventually come to accept this.[147]

The Biologic Institute statement resembles many issued by ID advocates, including the Discovery Institute and some of its fellows. The Biologic Institute, however, has fellows with seemingly solid scientific credentials, and this adds further force to the statement. As the preceding discussion explains, the only difference between a savvy marketing statement that is factually inaccurate and scientifically inane drafted on behalf of those with solid scientific credentials and one drafted on behalf of those without such credentials is that those with the credentials should know better. The reality is, however, that there are thousands, if not tens of thousands, of scientists with stellar credentials who reject design, and only a handful who seem to accept it, and they hide behind the marketing of ideas and proof by inference rather than the scientific method when it comes to their ultimate hypotheses.

Ironically, it is ID and creationist arguments that often confuse members of the public and slow the pace of such acceptance. From the perspective of ID advocates, this is a great achievement, and in a very real sense, it is. The achievement, however, is in the field of marketing, not of science.

B. Controversy (or not)

As the ID movement looks toward the future, the teach the controversy issue is likely to play a big role in both its legal and marketing strategies. This makes sense. Because courts are likely to find that ID is not science, as science is currently defined, and that it does have religious roots, the teach the controversy approach may be successful with a court that is predisposed to favor ID. As we saw in Chapter 3, it would take such a court for the argument to be successful because the teach the controversy approach would require a court to accept that a scientific controversy exists. This is because the teach the controversy approach is most often aimed at the science classroom. ID itself could be taught in philosophy or comparative religion courses so long as it is taught objectively. The raison d'être for the teach the controversy assertion is to support the notion that there is a scientific controversy to which people, especially students learning about evolution, should be exposed.

Think about this from a marketing perspective. Assume that a company has had little success breaking into a market of which it desperately wants to be a part. The company's product does not fit into that market as the market is currently structured. How might the company get access to the market? It might try what the excellent book on marketing strategy, *Marketing Warfare*, referred to as a *flanking tactic*.[148]

The book looks at the soda market to describe the tactics of variously sized companies in that market.[149] Here the situation is a little different because all the companies in the book, from Coca Cola to RC Cola, were already part of the soda market.[150] ID's very problem is claiming a place in the science market, but the science market has entrance rules that keep ID out. This is where the flanking tactic might be useful.[151] Rather than attempting to take the market leaders head-on – here the market leader is mainstream science, and a head-on approach would require that ID be recognized as science under the current standards for science – a product seeking access to a market (or seeking a greater market share) can flank,[152] that is, go around the strengths of the market leaders and attack from the sides.[153]

This is exactly what teach the controversy does. The teach the controversy strategy ignores the vast evidence supporting evolution and the fact that there is no controversy to mainstream scientists because the scientific evidence is so overwhelmingly supportive of evolution through natural selection. It simply proclaims that a controversy exists and challenges science to answer the call. In this way, it rhetorically places ID on the same playing field as mainstream science, even if in reality, science does not view it as such. The beauty to this approach is that the ID movement has flanked the scientific argument in the public mind, meaning that the fact that science or even the courts do not buy the existence of the controversy is irrelevant because inroads are made into the public consciousness, and this may affect future judges and politicians.

As we learned in Chapter 3, most courts are unlikely to view the teach the controversy argument as anything other than a religiously based argument against evolution or a backdoor attempt to get the ID concept, even if not the details, into the science classroom. Yet anyone who spends any time observing the behavior of courts, especially on religion issues, will come away from that experience with the understanding that one can usually expect the unexpected from some courts and judges. Depending on the evidence a court allows to come before it, the theory of the case set forth by the parties, and the predisposition of the judge or judges involved, a court might uphold the teach the controversy approach. This is not to say that such a holding would be consistent with the great weight of the evidence on this issue or with the governing law; rather it is only to say that some court somewhere might so hold.

Such a holding would add even more rhetorical fire to the teach the controversy argument. It could be used to disparage the decisions of courts that hold otherwise and to support the existence of a controversy. Nothing would change in the world of mainstream scientific research, at least for the time being, but the practical result would be a further public relations success in stacking the future deck in favor of ID. The fact that ID is still bad science in the eyes of the vast majority of scientists would be irrelevant to those who either do not know this or do not know that this is based on a staggering amount of evidence in favor of evolution through natural selection and the paucity of scientific evidence supporting ID.

This sort of legal tactic – finding a sympathetic court and stacking the deck in favor of religion in the public schools because excluding it would be discriminatory despite facts and precedent that point the other way – has been used before in the context of religion in the public schools with at least temporary success. ID advocates, of course, do not claim that ID is religious, but given its supernatural presumptions, it falls squarely within the Supreme Court's understanding of religion in establishment clause cases.[154] Moreover, the tactic described can, and has, worked in any number of legal contexts, whether the cases are focused on religion or not.

In 1993, shortly after the U.S. Supreme Court decided *Lee v. Weisman*,[155] the graduation prayer case that gave birth to the indirect coercion test discussed in Chapter 3, school boards all across the country received a letter from the American Center for Law and Justice (ACLJ).[156] The letter purported "to address the questions and concerns of school officials regarding the issues of prayer at graduation ceremonies."[157] However, the letter did so in a one-sided way. The *Lee* decision was portrayed as a limited decision, or at least not as the sound defeat for school prayer advocates that it appeared to be.[158] The letter suggested that organized student-initiated prayer would still be acceptable at school-sponsored graduations if it were voluntary and a majority of the graduating class voted for it.[159]

For those who understand American constitutional law, the majoritarian student-initiated prayer concept probably seems odd. After all, public school functions are still sponsored by the public schools, and organized prayer at such functions would have all the indicia of government sponsorship. The fact that a majority of students supported it would be irrelevant to a dissenter because the majority could only effectuate and enforce its will through mechanisms within the public school. If anything, the fact that one's peers voted for the prayer pursuant to a school policy requiring a vote on the issue might exacerbate a dissenter's sense of coercion and isolation. Beyond that, it is an old and strong constitutional principle that fundamental constitutional rights are not subject to a majority vote.[160]

It is likely that the ACLJ letter was based on an interpretation of *Lee* that focused on the role of the school principal in influencing and effectuating the prayer involved in that case. However, it is also significant that the *Lee* Court was quite concerned about the school district's role in sponsoring and controlling the graduation ceremony and that the opinion in *Lee* did not presume to set the outer boundaries of unconstitutional public school religious exercises. Thus the law set forth in earlier school prayer cases, such as *Abington Tp. v. Schempp*[161] and *Engel v. Vitale*,[162] including the holdings that constitutional rights are not subject to the whim of the majority, remained intact after *Lee*. In fact, *Lee* reinforced significant aspects of those cases.[163]

Significantly, a salient aspect of the student-initiated prayer concept as expressed in the ACLJ letter is the requirement that such prayer be voluntary. Given the language in *Lee* stating that a dissenter is essentially forced to protest or participate in the prayer ceremony, by at the very least remaining respectfully silent, and the *Lee* Court's focus on the inherent psychological and social pressure for children and adolescents to participate in religious exercises at public school functions, such prayer is not likely to be considered truly voluntary.[164] Of course, the student-initiated prayer concept attempts to vitiate this concern because it suggests that such prayer is not state action. Still, since the school district inevitably controls the forum for graduation ceremonies, one could argue that even if student-initiated prayer were not state action, the audience hearing it is still under the control of the state.

Yet not long after *Lee*, the Fifth Circuit Court of Appeals, in *Jones v. Clear Creek Independent School District*,[165] accepted the student-initiated prayer argument, so long as that prayer is voluntary, nonsectarian, and nonproselytizing. The *Jones* case originally arose before *Lee* was decided. Before the Supreme Court decided *Lee*, a panel of the Fifth Circuit Court of Appeals upheld a school policy created by the Clear Creek Independent School District in Texas that allowed prayer at high school graduations so long as a majority of the senior class voted to have it, a student delivered the prayer, and the prayer was "non-proselytizing and nonsectarian."[166] The court based this holding on its application of the *Lemon* test to the policy in question.[167] That decision has since become known as *Jones I*.[168]

Though the court's application of the *Lemon* test in *Jones I* was questionable, this soon appeared to be a moot point because the Supreme Court decided *Lee v. Weisman*, granted certiorari in *Jones*, vacated the *Jones* court's judgment, and remanded the case back to the Fifth Circuit for further consideration in light of *Lee*.[169] On remand, in *Jones II*, the Fifth Circuit panel again upheld the school policy, holding that it did not violate the indirect coercion test set forth in *Lee* or the endorsement test and that its earlier *Lemon* analysis was still applicable because the school board policy did not violate *Lee*.[170]

The *Jones II* opinion is written in a highly cynical tone in several places, and the court most definitely gives the impression that it disapproves of the Supreme Court's establishment clause jurisprudence – at least as it has traditionally been applied to the public schools. The *Jones II* court's analysis seems geared toward finding a way to uphold the graduation prayer in question despite serious questions raised by all three tests the court applies. This has caused other courts and scholars to criticize the opinion.[171]

The *Jones II* court held that under the facts of that case, the prayer in question does not violate the *Lemon* test because solemnization of the ceremony is a legitimate secular interest supporting ceremonial prayer; the primary effect of the prayer was solemnization of the ceremony and not the advancement of religion since it was nonsectarian; and there was no entanglement problem, even though the policy required the school to review the prayer to be sure it was nonsectarian, because there was no precedent that said that such yearly review was unconstitutional entanglement.[172] The court did not give much explanation as to why a religious means, prayer, was necessary to solemnize a graduation ceremony, nor did the court discuss why the ceremony would not be solemn without the prayer.

In stating that the nonsectarian nature of the prayer precluded the advancement of religion, the court failed to address the long line of Supreme Court precedent that holds that advancing religion in general over irreligion or vice versa can create an establishment clause violation. To an atheist or someone from a faith that finds public prayer offensive, the nonsectarian nature of the prayer would be irrelevant. In addressing the entanglement issue, the court only addressed the potential entanglement created by school board review of the prayer. The court did not even attempt to address the potential entanglement created by the fact that the board empowers the students to vote on the prayer through a formal policy or that the school controls the graduation ceremony.

The prayer did not violate the endorsement test according to the court because the policy left the decision whether there would be prayer to the students, and thus the court asserted that dissenting students would not perceive government endorsement.[173] Yet a reasonable observer familiar with the history of the policy and the apparent support for graduation prayer on the part of school officials as well as the fact that the policy empowers the majority to impose a religious exercise on the minority, or alternatively, to allow the majority to take such a religious exercise away from the minority in the unlikely event prayer was not favored by the majority would perceive endorsement.[174] It is hard to find a more effective method to create political insiders and outsiders along religious lines than to have a government entity set forth a policy that puts whether or not to have a religious exercise to a vote.

Still, the key issue the *Jones II* court had to face was whether the prayer in that case violated the coercion test set forth in *Lee*. In doing so, the *Jones II* court gleaned three requirements that must be met to establish a violation pursuant to *Lee*: (1) the government must direct (2) a formal religious exercise (3) in a manner that obliges the participation of objectors.[175] The court went on to hold that the school board does not direct the religious exercise because the school board policy permits the graduating class to determine whether prayer would occur and only allows a student to conduct such prayer.[176] Moreover, the court found no formal religious exercise because the policy simply tolerated the prayer without requiring or favoring it.[177] Finally, the court found that the policy did not coerce objectors to participate in the prayer because the policy allowed student participation in the prayer decision, and thus objecting students would be aware that the prayer represents the will of their peers and not of the state, especially since high school students are less impressionable than younger students.[178]

In a staggering, and given the tenor of the opinion, probably intentional oversight, the court does not address that the school board policy empowered the students to vote on the prayer, the prayer would occur at a school-controlled graduation, and the minority of students who oppose the prayer would be subject to the religious will of the majority as empowered by the state – factors certainly relevant to the state direction and coercion issues. Nor did the court analyze in detail that the *Lee* court held that high school students are subject to a great deal of peer pressure. Moreover, the court does not adequately explain why the prayer at Clear Creek was not a formal religious exercise, devoting six lines to this and citing nothing to support its conclusion.[179]

The court seems hell bent, if you will pardon the term, on finding the prayer to be constitutional and to voice its support for such religious exercises in the public schools, albeit indirectly, if not subtly. The *Jones II* decision was a serious victory for school prayer advocates at that time. School prayer advocates had been looking for a way to bring organized prayer back to the public schools since the 1962–1963 school prayer decisions, and even more passionately after *Lee* was decided.[180] The idea of voluntary, student-initiated prayer played well with much of the public, and it implied that any such prayer would perhaps be a form of free speech. To exclude it would be to discriminate against religion.[181] This argument was rejected by a number of the courts to address it,[182] limited to the facts of *Jones* by later panels of the U.S. Court of Appeals for the Fifth Circuit,[183] and ultimately rejected by the U.S. Supreme Court in *Santa Fe v. Doe*, albeit not in the graduation context.[184]

It is also worth mentioning here that the ACLJ letter implied that *Jones II* was controlling outside the Fifth Circuit.[185] At that time, I was working as an associate in a practice in the Mid-Atlantic that represented several school districts. One of the school superintendents who received the ACLJ letter was informed enough to realize that something was fishy and asked for a legal opinion. I was asked by a partner in the firm to research the issue. My conclusion was that the ACLJ letter was at the least an overstatement and at the most intellectual dishonesty or outright malpractice. The ACLJ letter did not clearly state that that organization's interpretation was only one possible interpretation of *Lee*, that *Jones* only governed in the Fifth Circuit because the denial of a petition for writ of certiorari has no precedential value and neither federal trial courts nor federal courts of appeals are bound by decisions in other circuits, and that it would likely cost a huge amount of money in litigation to find out if the *Jones II* court's decision would apply elsewhere.[186] The lawsuits following the letter, and the losses by several school boards, demonstrate that *Jones II* was essentially an outlier case decided by a court predisposed to uphold the prayer, yet you can see the legal and public relations mileage the *Jones v. Clear Creek* decision gave to school prayer advocates.

The ID movement can still realistically hope for such a victory and the marketing and legal mileage such a victory would gain for them if a court relied on the teach the controversy approach. Of course, the movement needs a court willing to ignore or limit the evidence and bend the analysis set forth by the Supreme Court in *Edwards* and *Epperson*.[187] It is also a given that such a court would have to ignore or distinguish *Kitzmiller*, but like the voluntary, student-initiated school prayer concept, teach the controversy may provide an excellent opportunity for a court to do so and the rhetorical basis for such a court to create at least a passable legal justification. The decision would be legally weak and factually limited, but that did not stop the *Jones II* court from deciding as it did.

Looking ahead, it is sensible to predict that teach the controversy, or some variation of it, will continue to be a major strategy used by the ID movement. A related question may be whether school boards will be the focus of this battle. As we saw in *Kitzmiller*, that may not be a winning strategy legally, and as we saw in Kansas, it may not be a winning strategy politically. Of course, every political situation is different, and a state to watch on the political front in the immediate future is Texas. The state school board there seems poised to be the next to step into the ID fray.[188] An alternative to a focus on school boards as the mechanism to promote ID or teach the controversy is a focus on teachers who want to teach it and are denied the ability to do so.

C. Have the teachers take the lead

The legal aspects of this were addressed in Chapters 3 and 6 and will not be revisited here. This section will provide a brief overview of the marketing potential for this approach. It will only be briefly discussed, however, because the legal arguments are so legally problematic that there is almost no chance of success even if a court predisposed to support ID were to hear the case.[189]

The benefits to having teachers raise the issue are several. First, when a teacher is denied the ability to teach ID (or the controversy), claims of discrimination and unfair exclusion will ring out.[190] This fits beautifully with the ID movement's general claims of discrimination and unfair exclusion. Though there is not likely to be a valid legal claim on behalf of the teacher, nor would mainstream scientists support such a teacher, a denial of this sort is a fund-raising bonanza. It also can raise the ire of many in the general public when it is spun to suggest that there was no valid basis for denying the teacher's request, as it would most likely be spun.

There is also image value in having a science teacher who wants to teach ID (or the controversy). This puts a real-life face on the dispute and may even provide a real-world voice for ID. This was done with solid marketing effect by a teacher in the Kansas hearings. It remains to be seen how many of these teachers will come forward and how many school boards will preclude them from teaching ID. It is a good bet that in many places, if a teacher does teach ID, complaints will be brought by some parents and the local scientific community. This may cause school boards to step in and prevent the teacher from teaching ID, or it may be the first time a school board becomes aware that one or more of its science teachers are doing so. Whereas some teachers may simply stop teaching ID when asked, in other cases, such a request or order by school officials is likely to lead the teacher to file grievances or even a lawsuit. As we learned in earlier chapters, the teacher will inevitably lose such a lawsuit, but the marketing benefit the ID movement gains will remain whether any lawsuit filed against the school board is won or lost.

Significantly, the ID movement will have gained another martyr in these circumstances, even if that martyr is an unwilling one. The marketing value of having a teacher's request or lesson plans rejected in this way, even if the decision by school officials is completely consistent with state and/or school district educational guidelines, would be high. Such a situation combines the potent marketing elements of discrimination, denial of academic freedom, and teach the controversy. These sorts of situations have arisen in recent years,[191] and it is likely that we will see more of them in the near future. There is no legal argument that is likely to allow the ID movement to win such a case in court, but as we have seen, the ID movement is as concerned about winning in the court of public opinion as it is about winning in a court of law, at least in the short term.

Conclusion

Many people in the legal, scientific, and academic worlds view the intelligent design (ID) movement as a group of deceptive and fringe fanatics. The ID movement should not be underestimated in this way. There is no question that the ID movement's so-called scientific arguments are exceptionally flawed and that when ID advocates play the scientific proof game, they are destined to lose. Of course, one of their goals is to change the rules of the game. This is why it is essential to understand what the ID movement is: a market-savvy movement grounded in religious apologetics. Much of the attention has focused on what the ID movement is not, namely, a respectable scientific movement.

The risk in underestimating the ID movement is huge because if it were to succeed in changing the rules of science, it would be hard to go back, and the pace of scientific advancement we take for granted today would slow dramatically. This book explains why changing the rules of science in the way the ID movement would need to do to be accepted as science would be damaging to science; would be inconsistent with current law; would cause educational anarchy; and ironically, would undermine the very worldview espoused by the ID movement. Few have focused on the role relativism must play in the ID movement's bid to change the rules of science, yet that role has significant scientific, legal, and philosophical import.

It is hard to understand the ID movement without considering the role religious apologetics played in forming the movement and the movement's approach. ID arose shortly after the U.S. Supreme Court decided *Edwards v. Aguillard*,[1] and much of its strategy seems aimed to avoid similar legal defeats. Most notable in this regard is ID's failure to name the designer. Even though virtually every shred of evidence on the origin, strategy, and views of ID and ID advocates points toward the designer being God, the movement learned from the defeat of creation science in *Edwards* to avoid openly acknowledging its religious motivations.[2]

This, along with the rhetorical arguments and "scientific" exploits of ID advocates, has caused some to accuse the movement of dishonesty. The obvious import is that a religiously grounded movement, or one that at least openly acknowledges its support for the concept of a strong values system, engages in inexcusable hypocrisy by being dishonest. As we saw earlier, however, this assertion is question begging. Religious apologists often believe the ends (the service to God) justify the rhetorical means, and therefore higher values may demand a certain amount of spin to convince people of the so-called truth. I believe this accounts for the ID movement's savvy marketing plan that seems to the scientific and legal world to promote intellectual dishonesty at the very least, if not outright bunk.

Those in the ID movement probably do not view ID as dishonest and almost certainly believe that it serves a higher purpose. That purpose is most likely sectarian for the bulk of ID advocates, but for some, it may truly be about the exclusion of supernatural causation generally from mainstream science without reference to a specific God. One would have to enter the hearts and minds of ID advocates to know who truly views ID as serving sectarian ends, although a number of public statements by ID advocates, along with the Wedge Document, suggest that sectarian concerns may influence many in the ID movement.

This book has explained the legal, scientific, and philosophical implications that arise when we consider what the ID movement is really saying. The arguments made by ID advocates directly implicate scientific relativism, and as we have seen, ID is ill equipped to take advantage of relativism of any sort. But this assumes that we seriously consider the implications and context underlying what ID advocates actually say and do. In taking the implications and context seriously, we often miss what is really going on. ID is a marketing strategy, and from a marketing perspective, implications and context can be spun out of the picture to focus on the message, whether it be "teach the controversy" or claims of intellectual discrimination. When we understand that ID is a well-funded, well-run movement that is marketing creation, we can begin to contemplate how to address it, and we can begin to understand why it has not, and will not, go away, despite the legal and scientific defeats it has suffered.

This book has shown that one can only accept the ID paradigm if one accepts ID's precommitment to a particular conclusion, namely, that complex natural phenomena were designed by a supernatural being or beings. Mainstream science, to the extent it relies on a precommitment to any particular paradigm, relies on a methodological precommitment that is designed to test and/or falsify particular conclusions and hypotheses. These are very different sorts of precommitments. ID presumes a substantive end – the existence of a supernatural designer – and will not abandon or try to disprove this end, whereas mainstream science adapts to the evidence, and when enough scientific evidence is mounted using the scientific method, science abandons hypotheses that have been disproved. Given the overwhelming amount of scientific evidence supporting evolution through natural selection, there is no scientific basis to abandon evolution. Perhaps there is a religious basis, but if so, ID advocates will never acknowledge it.

Notes

1. Designing Design

1. Excerpt from Discovery Institute, Wedge Document (1998), an internal memorandum since released to the general public on the Web. The full text of the Wedge Document is available at http://www.antievolution.org/features/wedge.html. The Wedge Document is discussed in detail in BARBARA FORREST & PAUL R. GROSS, CREATIONISM'S TROJAN HORSE (2004).
2. Edwards v. Aguillard, 482 U.S. 578, 599 (1987) (Justice Powell concurring), citing Malnak v. Yogi, 592 F.2d 197 (3d Cir. 1979).
3. Edwards, 482 U.S. 578; Epperson v. Arkansas, 373 U.S. 97 (1968).
4. Edwards, 482 U.S. 578; Epperson, 373 U.S. 97.
5. *Cf.* PHILLIP JOHNSON, DARWIN ON TRIAL (1991) (discussing scientific materialism); Wedge Document, *supra* note 1 (same).
6. *See generally* FORREST & GROSS, *supra* note 1 (discussing the strategy of ID advocates to gain public acceptance by using scientific jargon to mask their religious base); ROBERT PENNOCK, TOWER OF BABEL: THE EVIDENCE AGAINST THE NEW CREATIONISM (1999) (same, and also discussing the evolution of ID from

earlier forms of creationism); *see also* Kent Greenawalt, *Establishing Religious Ideas: Evolution, Creationism, and Intelligent Design*, 17 NOTRE DAME J.L. ETHICS & PUB. POL'Y 321 (2003) (thoughtful discussion about ID in public schools and the broader scientific and philosophical questions raised in that context).

7. Ironically, many biblical creationists could find ID troubling because it denies what they believe to be biblically mandated truth by failing to acknowledge openly that the designer is God and because it seemingly rejects certain literalistic interpretations of the book of Genesis. *Cf.* Pennock, *supra* note 6, at 18–26, 226–28 (discussing similar battles between old Earth creationists and young Earth creationists).
8. In fact, for over a thousand years, some Jewish and Christian theologians have acknowledged similar ideas, and the Roman Catholic Church has recently acknowledged this position. John Thavis, *Evolution and Creation: A Recurring Papal Theme, Often Misunderstood*, CATHOLIC NEWS SERVICE, Feb. 1, 2008.
9. WILLIAM A. DEMBSKI, INTELLIGENT DESIGN: THE BRIDGE BETWEEN SCIENCE & THEOLOGY (1999); MICHAEL BEHE, DARWIN'S BLACK BOX (1998).
10. Frank S. Ravitch, *Playing the Proof Game: Intelligent Design and the Law*, 113 PENN ST. L. REV. 841 (2009).
11. *See infra* notes 180–82, and accompanying text.
12. PENNOCK, *supra* note 6, at 344–77; *see generally* FORREST & GROSS, *supra* note 1 (discussing the marketing strategies of the ID movement).
13. Kitzmiller v. Dover Area School District, 400 F.Supp.2d 707, 735–46 (E.D. Pa. 2005).
14. *Cf.* David K. DeWolf et al., *Teaching the Origins Controversy: Science, or Religion, or Speech?* 2000 UTAH L. REV. 39, 106 (2000) ("While public schools are not public fora per se, they are publicly funded places where ideas are exchanged. Thus, if public schools or other governmental agencies bar teachers from teaching about design theory but allow teachers to teach neo-Darwinism, they will undermine free speech and foster viewpoint discrimination"); *id.* at 56–57 ("Thus, those biologists who seek to insulate their

preferred theories from critique by rhetorical gerrymandering – that is, by equating dominant evolutionary theories with science itself and then treating all criticism of such theories as necessarily "unscientific" – themselves act in a profoundly unscientific manner").

15. *Id.*; Francis J. Beckwith, *Public Education, Religious Establishment, and the Challenge of Intelligent Design*, 17 NOTRE DAME J.L. ETHICS & PUB. POL'Y 461, 489–90 (2003) ("Thus, forbidding the teaching of ID [or legitimate criticisms of evolution] in public schools because it lends support to a religion, while exclusively permitting or requiring the teaching of evolution, might be construed by a court as viewpoint discrimination, a violation of state neutrality on matters of religion, and/or the institutionalizing of a metaphysical orthodoxy, for ID and evolution are not two different subjects [the first religion, the second science] but two different answers about the same subject"); DeWolf et al., *supra* note 14, at 58 ("But clearly students would not be well served by presenting a false picture of agreement where in fact there is controversy").
16. *See* Jay D. Wexler, *The Scopes Trope*, 93 GEO. L.J. 1693, 1695–96 (2005) (reviewing LARRY A. WITHAM, WHERE DARWIN MEETS THE BIBLE [2002]) ("intelligent design advocates have argued that notions of academic freedom, equality, and educational comprehensiveness require school boards and officials to allow teachers to introduce students to intelligent-design theory and, in some cases, even require them to do so").
17. Ravitch, *supra* note 10, at 881–82, 884–85.
18. *Id.* at 874–85.
19. *Kitzmiller*, 400 F.Supp.2d, 735–46.
20. *Id.* at 736.
21. *See* FORREST & GROSS, *supra* note 1 (discussing the Wedge Document developed by leading ID advocates that sets forth these concerns). The full text of the Wedge Document can be found *infra* note 130.
22. *Id.*
23. FRANCIS J. BECKWITH, LAW, DARWINISM, AND PUBLIC EDUCATION (2003).

24. JOHNSON, *supra* note 5.
25. FRANK S. RAVITCH, MASTERS OF ILLUSION: THE SUPREME COURT AND THE RELIGION CLAUSES (2007), at 109–10.
26. JORN RUSSEN & HENNER LAASS, HUMANISM IN INTERCULTURAL PERSPECTIVE (2010).
27. PAUL KURTZ ed., HUMANIST MANIFESTOS I AND II (1973).
28. The interaction between science and morality will be explored in greater depth in Chapters 4 and 5.
29. RICHARD DAWKINS, THE GOD DELUSION (2006).
30. DANIEL C. DENNETT, DARWIN'S DANGEROUS IDEA (1995); Richard Lewontin, *A Review of the Demon-Haunted World: Science as a Candle in the Dark*, NEW YORK REV. OF BOOKS, Jan. 9, 1997.
31. *See supra* notes 29–30, and accompanying text; *see also* Chapter 5.
32. *See* Chapter 4.
33. KENNETH R. MILLER, FINDING DARWIN'S GOD (1999).
34. *See* Wedge Document, *supra* note 1; JOHNSON, *supra* note 5.
35. *See supra* notes 29–30, and accompanying text; *see also* Chapter 5.
36. MILLER, *supra* note 33.
37. *Id.*
38. *Id.*
39. *Id.* at 239–43; DAWKINS, *supra* note 29.
40. EXPELLED: NO INTELLIGENCE ALLOWED (Premise Media Corp. 2008).
41. *Id.*
42. *See* Chapter 6.
43. *Id.*
44. *See infra* Chapter Six.
45. Edwards v. Aguillard, 482 U.S. 578, 587 (1987).
46. *Id.* at 586.
47. *Id.* at 581, 586.
48. *Id.* at 587.
49. *Id.* at 586–89.
50. ROBERT PENNOCK, TOWER OF BABEL: THE EVIDENCE AGAINST THE NEW CREATIONISM (1999).

51. *Id.* at 10–26; RONALD L. NUMBERS, THE CREATIONISTS: FROM SCIENTIFIC CREATIONISM TO INTELLIGENT DESIGN (expanded ed. 2006).
52. NUMBERS, *supra* note 51.
53. The specific date and year for creation as accepted by many biblically literalist Protestant sects was estimated to be October 23, 4004 BCE, by Dr. John Lightfoot, vice chancellor of Cambridge University in 1644. Over a decade later, in 1658, Bishop James Ussher, an Anglican clergyman and vice chancellor of Trinity College in Ireland, published the book THE ANNALS OF THE WORLD, upon which many young Earth creationists rely for the date of creation. Ussher's year and date for the beginning of creation are identical to Lightfoot's (although there was a discrepancy about the exact hour of creation, which Ussher did not include in his account).
54. NUMBERS, *supra* note 51.
55. FORREST & GROSS, *supra* note 1.
56. Edwards, 482 U.S. 578; *see also* Daniel v. Waters, 515 F.2d 485 (6th Cir. 1975) (finding a Tennessee statute requiring "balanced treatment" in textbooks unconstitutional); McLean v. Arkansas Bd. of Educ., 529 F.Supp. 1255 (E.D. Ark. 1982) (same for Arkansas balanced treatment act).
57. Edwards, 482 U.S. at 596.
58. *Id.* at 581–82.
59. 482 U.S. 578 (1987).
60. *Id.* at 586–89, 591–93, 596–97.
61. *Id.*
62. *Id.* at 581, 586–89.
63. *Id.* at 587, 591–93.
64. *Id.* at 591.
65. *Id.* at 591 note 13.
66. *Id.* at 592–94, 596–97.
67. *Id.* at 586–88.
68. *Cf. id.* at 587–89 ("under the Act's requirements, teachers who were once free to teach any and all facets of this subject are now unable to do so").

69. *Id.* at 587–94.
70. *Id.* at 589.
71. *Id.* at 588–89, 592–93.
72. *Id.* at 588.
73. *Id.*
74. *Id.* at 590–94; *id.* at 599 (Justice Powell concurring).
75. *Id.* at 591.
76. Edwards, 482 U.S. 578; Waters, 515 F.2d 485 (finding Tennessee balanced treatment law unconstitutional); McLean, 529 F.Supp. 1255 (same for Arkansas act).
77. FORREST & GROSS, *supra* note 1.
78. *Id.*
79. *Id.* at 217–39; *see also* Beckwith, *supra* note 15 (arguing, in part, that ID can and should be taught in public schools).
80. See DeWolf et al., *supra* note 14.
81. PENNOCK, *supra* note 6.
82. 393 U.S. 97 (1968).
83. *Id.*
84. *Id.* at 98–99.
85. *Id.* at 99.
86. *Id.* at 100.
87. *Id.*
88. *Id.* at 109.
89. *Id.* at 107–10.
90. *Id.* at 107.
91. *Id.* at 103–04, 109.
92. *See, e.g.*, Sch. Dist. of Abington Tp. v. Schempp, 374 U.S. 203 (1963) (applying neutrality concept to find school prayer and Bible reading unconstitutional); Zelman v. Simmons-Harris, 536 U.S. 639 (2002) (applying formal neutrality to uphold school voucher program); *see also* Douglas Laycock, *Formal, Substantive, and Disaggregated Neutrality toward Religion*, 39 DEPAUL L. REV. 993 (1990) (analyzing formal and substantive neutrality and rejecting formal neutrality); Daniel O. Conkle, *The Path of American Religious Liberty: From the Original Theology to Formal Neutrality and an Uncertain Future*, 75 IND. L. J. 1, 8–10 (2000)

(discussing formal neutrality and the trend toward its increased use by the Court); Frank S. Ravitch, *A Funny Thing Happened on the Way to Neutrality: Broad Principles, Formalism and the Establishment Clause*, 38 GA. L. REV. 489 (2004) (acknowledging that courts have regularly used a variety of neutrality concepts in deciding cases but arguing that there is no neutral place from which one can say that a given neutrality approach is neutral).

93. Epperson, 393 U.S. at 107–09.
94. *Id.* at 98–99.
95. *Id.* at 109.
96. *See generally* PENNOCK, *supra* note 6 (discussing the evolution of the creation science movement from factions within the broader creationism movement).
97. *See* Chapter 3.
98. PENNOCK, *supra* note 6.
99. *See* PENNOCK, *supra* note 6; NUMBERS, *supra* note 51; Ravitch, *supra* note 10.
100. *Id.* (all sources).
101. Kitzmiller, 400 F.Supp.2d 707.
102. *See infra* notes 237–48; *see also* Chapter 7.
103. *Id.*
104. FORREST & GROSS, *supra* note 1.
105. *Id.* at 275–76.
106. *Id*; Kitzmiller v. Dover Area School District, 400 F.Supp.2d 707, 716–23, 735–46 (E.D. Pa. 2005); *see also* Richard B. Katskee, *Why It Mattered to Dover That Intelligent Design Isn't Science*, 5 FIRST AMEND. L. REV. 112, 119 (2006) (noting that ID is in part a response to the *Edwards* decision); Nicholas A. Schuneman, *One Nation Under... The Watchmaker? Intelligent Design and the Establishment Clause*, 22 BYU J. PUB. L. 179, 186 (2007) (same); Kevin Trowel, Note, *Divided by Design: Kitzmiller v. Dover Area School District, Intelligent Design, and Civic Education*, 95 GEO. L.J. 855, 858 (2007) (same).
107. *See generally* FORREST & GROSS, *supra* note 1 (setting forth ID movement's strategy to claim the mantle of science and gain

public recognition); DEMBSKI, *supra* note 9 (early work by leading ID advocate making such arguments).

108. *Supra* note 40.
109. *See* Chapters 2 and 7.
110. For an example of Christian apologetics on a wide range of topics, including creationism, *see* KREEFT & TACELLI, HANDBOOK OF CHRISTIAN APOLOGETICS (1994).
111. *Cf. id.* (taking beliefs underlying apologist arguments, and those arguments themselves, as having important meaning and correctness); *see also* Chapter 7.
112. *See* Wedge Document, *supra* note 1; JOHNSON, *supra* note 5.
113. FORREST & GROSS, *supra* note 1 at 15–20, 174.
114. *Id.* at 18–19.
115. JOHNSON, *supra* note 5.
116. FORREST & GROSS, *supra* note 1, at 15–23.
117. This definition is from the Discovery Institute's Center for the Renewal of Science and Culture Web site as it originally existed in 1996. The language on that site has since been changed (although links to the original site are widely available on the Web), but this or very similar language is found in the Discovery Institute's Wedge Document, the writings of numerous intelligent design writers, and Ben Stein's recent movie *Expelled: No Intelligence Allowed* (2008).
118. JOHNSON, *supra* note 5.
119. FORREST & GROSS, *supra* note 1.
120. JOHNSON, *supra* note 5; *infra* note 130.
121. *Id.*
122. *Id.*
123. Stephen Jay Gould, *Impeaching a Self Appointed Judge*, 267 SCIENTIFIC AMERICAN 118–121 (July 1992).
124. FORREST & GROSS, *supra* note 1, at 15–23.
125. *Id.* at 19–23 and generally.
126. *Id.* at 15–23.
127. *See generally id.* (providing detailed discussion of the wedge strategy and the Wedge Document).
128. *Id.*

129. *Id.*; *see also infra* note 130.
130. The full version of the Wedge Document is readily available on the Web and in several texts. It reads in part as follows:

INTRODUCTION
The proposition that human beings are created in the image of God is one of the bedrock principles on which Western civilization was built. Its influence can be detected in most, if not all, of the West's greatest achievements, including representative democracy, human rights, free enterprise, and progress in the arts and sciences.

Yet a little over a century ago, this cardinal idea came under wholesale attack by intellectuals drawing on the discoveries of modern science. Debunking the traditional conceptions of both God and man, thinkers such as Charles Darwin, Karl Marx, and Sigmund Freud portrayed humans not as moral and spiritual beings, but as animals or machines who inhabited a universe ruled by purely impersonal forces and whose behavior and very thoughts were dictated by the unbending forces of biology, chemistry, and environment. This materialistic conception of reality eventually infected virtually every area of our culture, from politics and economics to literature and art.

***

Discovery Institute's Center for the Renewal of Science and Culture seeks nothing less than the overthrow of materialism and its cultural legacies. . .

THE WEDGE STRATEGY
FIVE YEAR STRATEGIC PLAN SUMMARY
The social consequences of materialism have been devastating. As symptoms, those consequences are certainly worth treating. However, we are convinced that in order to defeat materialism, we must cut it off at its source. That source is scientific materialism. This is precisely our strategy. If we view the predominant materialistic science as a giant tree, our strategy is intended to function as a "wedge" that, while relatively small, can split the trunk when applied at its weakest points. The very beginning of this strategy, the "thin edge of the wedge," was Phillip Johnson's critique of Darwinism begun in 1991 in Darwinism on Trial, and continued in Reason in the Balance and Defeating Darwinism by Opening Minds. Michael Behe's

highly successful Darwin's Black Box followed Johnson's work. We are building on this momentum, broadening the wedge with a positive scientific alternative to materialistic scientific theories, which has come to be called the theory of intelligent design (ID). Design theory promises to reverse the stifling dominance of the materialist worldview, and to replace it with a science consonant with Christian and theistic convictions.

The Wedge strategy can be divided into three distinct but interdependent phases, which are roughly but not strictly chronological. We believe that, with adequate support, we can accomplish many of the objectives of Phases I and II in the next five years (1999–2003), and begin Phase III (See "Goals/Five Year Objectives/Activities").

Phase I: Research, Writing and Publication
Phase II: Publicity and Opinion-making
Phase III: Cultural Confrontation and Renewal

***

Phase II. The primary purpose of Phase II is to prepare the popular reception of our ideas. The best and truest research can languish unread and unused unless it is properly publicized. For this reason we seek to cultivate and convince influential individuals in print and broadcast media, as well as think tank leaders, scientists and academics, congressional staff, talk show hosts, college and seminary presidents and faculty, future talent and potential academic allies. Because of his long tenure in politics, journalism and public policy, Discovery President Bruce Chapman brings to the project rare knowledge and acquaintance of key op-ed writers, journalists, and political leaders. This combination of scientific and scholarly expertise and media and political connections makes the Wedge unique, and also prevents it from being "merely academic."... Alongside a focus on influential opinion-makers, we also seek to build up a popular base of support among our natural constituency, namely, Christians. We will do this primarily through apologetics seminars. We intend these to encourage and equip believers with new scientific evidences that support the faith, as well as to "popularize" our ideas in the broader culture.

Phase III. Once our research and writing have had time to mature, and the public prepared for the reception of design theory, we will move toward direct confrontation with the advocates of materialist science through challenge conferences in significant academic settings. We will also pursue possible legal assistance in response to resistance to the integration of design theory into public school science curricula....

GOALS
* To defeat scientific materialism and its destructive moral, cultural and political legacies.
* To replace materialistic explanations with the theistic understanding that nature and human beings are created by God.

131. FORREST & GROSS, *supra* note 1, at 25.
132. These attempts have been unsuccessful in court so far. Kitzmiller v. Dover Area School District, 400 F.Supp.2d 707, 718–19 (E.D. Pa. 2005).
133. *Id.*; *infra* Chapter 3.
134. *See generally* Kitzmiller, 400 F.Supp.2d 707 (using the Wedge Document as part of the analysis of whether ID is scientific and/or religious and concluding that it is not scientific and is religiously grounded).
135. *See infra* notes 180–82, and accompanying text.
136. Edwards, 482 U.S. 578; Kitzmiller, 400 F.Supp.2d 707.
137. *See infra* notes 183–86, and accompanying text.
138. *See* Wedge Document, *supra* note 130.
139. *Id.*; *see also* JOHNSON, *supra* note 5.
140. Edwards, 482 U.S. 578; Kitzmiller, 400 F.Supp.2d 707.
141. JOHNSON, *supra* note 5; BEHE, *supra* note 9; DEMBSKI, *supra* note 9.
142. BEHE, *supra* note 9; DEMBSKI, *supra* note 9.
143. *Id.*; Wedge Document, *supra* note 130.
144. ID proponents are not alone in couching naturalism in these sorts of terms. A leading opponent of ID and supporter of this view of naturalism has made similar arguments to ID proponents on the meaning of scientific materialism and naturalism. *See* RICHARD DAWKINS, RIVER OUT OF EDEN 132–33 (1995) ("The Universe we observe has precisely the properties we should expect if there is,

at bottom, no design, no purpose, no evil and no good, nothing but blind, pitiless indifference").

145. *Cf.* Wedge Document, *supra* note 130; JOHNSON, *supra* note 5.
146. Michael Behe acknowledged this while on the stand during the *Kitzmiller* trial, when he acknowledged that the definition of science would have to be changed to include ID and that astrology would count as science under the new definition that would include ID. Kitzmiller, at 736.
147. Wedge Document, *supra* note 130.
148. MILLER, *supra* note 33.
149. *Id.*
150. *Id.*
151. *Id.*
152. *See generally id.* (the notion that these natural mechanisms are the work of God is a religious question and not a scientific one, but the latter approach does not preclude the former belief).
153. BEHE, *supra* note 9; DEMBSKI, *supra* note 9; JOHNSON, *supra* note 5; Wedge Document, *supra* 130.
154. Jay D. Wexler, *Darwin, Design, and Disestablishment: Teaching the Evolution Controversy in Public Schools*, 56 VAND. L. REV. 751, 814 (2003).
155. Kitzmiller, 400 F.Supp.2d at 718–20; FORREST & GROSS, *supra* note 1, at 15–23.
156. *See id.* (both sources).
157. BEHE, *supra* note 9; Beckwith, *supra* note 15; DEMBSKI, *supra* note 9; JOHNSON, *supra* note 5.
158. *See, e.g.*, ROBERT INKPEN, SCIENCE, PHILOSOPHY, AND PHYSICAL GEOGRAPHY (2005); HENRY PERKINSON, FLIGHT FROM FALLIABILITY: HOW THEORY TRIUMPHED OVER EXPERIENCE IN THE WEST (2001); Fouad Abd-El-Khalick, *On the Role and Use of "Theory,"* 91 SCIENCE EDUCATION 187 (2006); Ken Friedman, *Theory Construction in Design Research: Criteria: Approaches and Methods*, 24 DESIGN STUDIES 507 (2003); Simon Stow, *Theoretical Downsizing and the Lost Art of Listening*, 28 PHILOSOPHY AND LITERATURE 192 (2004).

159. BEHE, *supra* note 9; Beckwith, *supra* note 15; DEMBSKI, *supra* note 9; JOHNSON, *supra* note 5.
160. *See supra* note 158, and accompanying text.
161. *Cf.* MILLER, *supra* note 33, at 54 ("Evolution is *both* fact and theory. It is a fact that evolutionary change took place. And evolution is also a theory that seeks to explain the detailed mechanism behind that change").
162. *Id.* at 36–56.
163. *Id.* at 54.
164. *Id.*; KARL POPPER, LOGIC OF SCIENTIFIC DISCOVERY (1959).
165. POPPER, *supra* note 164; *cf. generally* LARRY LAUDAN, SCIENCE AND RELATIVISM (1990) (dialogue between positivist, realist, pragmatist, and relativist, where each recognizes that this occurs as a practical matter in "normal" science but disagree on the epistemological question of whether one can know that any scientific paradigm is better than another).
166. POPPER, *supra* note 164.
167. THOMAS KUHN, THE STRUCTURE OF SCIENTIFIC REVOLUTIONS (3rd ed. 1996) (recounting historical "scientific" frameworks based on anecdotal or purely deductive arguments).
168. S. N. Shore, *The Milky Way: Four Centuries of Discovery of the Galaxy, in* HOW DOES THE GALAXY WORK? A GALACTIC TERTULIA WITH DON COX AND RON REYNOLDS 1 (Emilio J. Alfaro et al. eds., 2004) ("Recognizing the Milky Way as one of a vast number of stellar systems was one product of the last century, and the birth of modern observational cosmology").
169. WILLIAM K. HARTMANN & CHRIS IMPEY, ASTRONOMY: THE COSMIC JOURNEY 424 (6th ed. 2002) ("Galaxies are not randomly distributed through space; they tend to cluster in different-size groups. Most of the galaxies in the Local Group are clumped into two subgroups, those around the Milky Way and those around the Andromeda galaxy"); NAT'L RESEARCH COUNCIL, THE DECADE OF DISCOVERY IN ASTRONOMY AND ASTROPHYSICS 47 (1991) ("Groupings of galaxies, which astronomers call 'structures,' are intriguing in their own right. Structures of various sizes abound, and

astronomers want to understand the nature of these structures and how they were born").

170. *See* HARTMANN & IMPEY, *supra* note 169, at 487 ("Galaxy formation is one of the most hotly debated topics in modern cosmology").
171. GUILLERMO GONZALEZ & JAY W. RICHARDS, THE PRIVILEGED PLANET x (2004) ("The fact that our atmosphere is clear; that our moon is just the right size and distance from Earth, and that its gravity stabilizes Earth's rotation; that our position in the galaxy is just so; that our sun is its precise mass and composition – all of the facts and many more not only are necessary for Earth's habitability but also have been surprisingly crucial to the discovery and measurement of the universe by scientists").
172. PENNOCK, *supra* note 6.
173. FORREST & GROSS, *supra* note 1; MILLER, *supra* note 33, at 126–28.
174. MILLER, *supra* note 33, at 126–28.
175. WILLIAM PALEY, NATURAL THEOLOGY (Bobbs-Merrill paperback ed. 1963).
176. *Id.*
177. ALISTER MCGRATH, THE ORDER OF THINGS: EXPLORATIONS IN SCIENTIFIC THEOLOGY (2006); Stephanie L. Shemin, *The Potential Constitutionality of Intelligent Design*, 13 GEORGE MASON L. REV. 621, 663–64 (2005).
178. PALEY, *supra* note 175.
179. *Id.*
180. PLATO, TIMAEUS (Peter Kalkavage trans., 2001).
181. CICERO, DE NATURA DEORUM I (Richard McKirahan ed., 1997).
182. THOMAS AQUINAS, THE SUMMA THEOLOGICA OF ST. THOMAS AQUINAS (1981).
183. PALEY, *supra* note 175.
184. NORMAN L. GEISLER, CHRISTIAN APOLOGETICS (1988).
185. PALEY, *supra* note 175.
186. CHARLES DARWIN, THE ORIGIN OF SPECIES BY MEANS OF NATURAL SELECTION (1999) (reprint of the first edition published in 1859).
187. *Cf.* PALEY, *supra* note 175, *with* DARWIN, *supra* note 186 (Darwin was clearly influenced by the observations of earlier naturalists

such as Paley but not by the theological aspects of that naturalism).

188. DARWIN, *supra* note 186.
189. *See generally id.*
190. PENNOCK, *supra* note 6.
191. *See generally id.* (setting forth an exceptionally well detailed account of the relationship between natural theology, creationism, creation science, and ID).
192. BEHE, *supra* note 9.
193. *Id.* at 39.
194. *Id.*
195. *Id.*
196. *Id.*
197. PENNOCK, *supra* note 6, at 267–68.
198. *Id.*
199. Kenneth R. Miller, *The Flaw in the Mousetrap*, NATURAL HISTORY 75 (April 2002); Kenneth R. Miller, *The Moustrap Analogy or Trapped by Design*, http://www.millerandlevine.com/km/evol/DI/Mousetrap.html (last visited on Aug. 20, 2008).
200. Keith Robison, *Darwin's Black Box Irreducible Complexity or Irreproducible Irreducibility?*, http://www.talkorigins.org/faqs/behe/review.html.
201. MILLER, *supra* note 33, at 129–64.
202. *Id.* at 149–50.
203. *Id.* at 145–63.
204. *Id.* at 152–58.
205. BEHE, *supra* note 9.
206. MILLER, *supra* note 33, at 142–43.
207. Pallen, Penn, and Chaudhuri, *Bacterial Flagellar Diversity in the Post-genomic Era*, 13 TRENDS IN MICROBIOLOGY 143 (2005); Pallen, Beatson, and Bailey, *Bioinformatics, Genomics and Evolution of Non-flagellar Type III Secretion Systems: A Darwinian Perspective*, 29 FEMS MICROBIOLOGY REV. 201 (2005).
208. *See supra* notes 197–207, and accompanying text.
209. *Id.*
210. PENNOCK, *supra* note 6, at 263–72.

211. BEHE, *supra* note 9; PENNOCK, *supra* note 6, at 263–72.
212. PENNOCK, *supra* note 6.
213. *Id.* at 263–72.
214. DEMBSKI, *supra* note 9.
215. *Id.*
216. *Id.*
217. *Id.* at 128.
218. Wesley Elsberry and Jeffrey Shallit, *Playing Games with Probability: Dembski's Complex Specified Information*, *in* WHY INTELLIGENT DESIGN FAILS: A SCIENTIFIC CRITIQUE OF THE NEW CREATIONISM (2004).
219. *Id.*; DEMBSKI, *supra* note 9.
220. *Id.*
221. *See infra* notes 192–220, and accompanying text.
222. *See infra* notes 237–48, and accompanying text; *see also* Chapter 7.
223. PENNOCK, *supra* note 6, at 163–72; MILLER, *supra* note 33, at 54–56.
224. PENNOCK, *supra* note 6, at 163–72.
225. MILLER, *supra* note 33, at 54–56.
226. *Id.*
227. *See generally* PENNOCK, *supra* note 6 (Pennock repeatedly points out the flaws in ID and creationist deductive arguments about the meaning of gaps in scientific theory as well as numerous other flaws in ID and creationist deductive reasoning).
228. *See supra* notes 192–213, and accompanying text.
229. *See supra* notes 214–20, and accompanying text.
230. For information about and the history of the "Church of the Flying Spaghetti Monster," *see* http://www.venganza.org. Suffice it to say that this "church" is a hilarious parody of, and commonsense response to, some of the ID movement's more outrageous claims.
231. *See supra* notes 192–213, and accompanying text.
232. MILLER, *supra* note 33, at 126–28.
233. *Id.*
234. *See generally id.*
235. PENNOCK, *supra* note 6, at 10–26.

236. *Id.*
237. FORREST & GROSS, *supra* note 1, at 206, 235–36, 337.
238. *Id.*; Wedge Document, *supra* note 1.
239. *Id.*
240. *Supra* note 40.
241. *See generally* PENNOCK, *supra* note 6 (pointing out that ID is not serious competition to mainstream science and is not perceived by scientists as such); MILLER, *supra* note 33 (same).
242. This is a reference to the band Tesla, not to Nikolai Tesla.
243. Cornelia Dean, *Scientists Feel Miscast in Film on Life's Origin*, N.Y. TIMES, Sept. 27, 2007, http://www.nytimes.com/2007/09/27expelled.html.
244. FORREST & GROSS, *supra* note 1; MILLER, *supra* note 33; PENNOCK, *supra* note 6.
245. Ellen Lloyd, *Genetic Engineering in Ancient Times*, AMERICAN CHRONICLE (October 12, 2006), http://www.americanchronicle.com/articles/14765; Archeology, Astronautics, and SETI Research Association, http://legendarytimes.com/index.php?op=page&pid=1 (last visited May 19, 2008).
246. Robert Roy Britt, *How Comets Might Seed Planets*, SPACE.COM (Jan. 24, 2000), http://www.space.com/scienceastronomy/planetearth/dna_spacegerm_000124.html.
247. Lloyd, *supra* note 245.

2. Big D and Manufactured Paradigm Shifts

1. *See* Kitzmiller, 400 F.Supp.2d, at 736–37; ROBERT PENNOCK, TOWER OF BABEL: THE EVIDENCE AGAINST THE NEW CREATIONISM (1999); Discovery Institute, Center for Science and Culture, Academic Freedom page, http://www.discovery.org/csc/freeSpeechEvolCampMain.php.
2. KARL POPPER, LOGIC OF SCIENTIFIC DISCOVERY (1959).
3. *Id.*
4. THOMAS KUHN, THE STRUCTURE OF SCIENTIFIC REVOLUTIONS (3rd ed. 1996).
5. *Id.*

6. Frank S. Ravitch, *Intelligent Design in Public University Science Departments: Academic Freedom or Establishment of Religion*, 16 WM. & MARY BILL RTS. J. 1061 (2008).
7. Kitzmiller, 400 F.Supp. at 735–46.
8. KUHN, *supra* note 4 (discussing paradigms in the sciences and asserting that there is no superparadigm to decide between conflicting paradigms).
9. *Id.*
10. *Id.*
11. *Id.* at 167–69, 177–78, 294.
12. PHILLIP JOHNSON, DARWIN ON TRIAL (1991); Discovery Institute, Wedge Document (1998), http://www.antievolution.org/features/wedge.html.
13. David K. DeWolf, Stephen C. Meyer, & Mark Edward DeForrest, *Teaching the Origins Controversy: Science, or Religion, or Speech?*, 2000 UTAH L. REV. 39, 68–75 (2000); Jay D. Wexler, Note, *Of Pandas, People, and the First Amendment: The Constitutionality of Teaching Intelligent Design in the Public Schools*, 49 STAN. L. REV. 439, 466–67 (1991) (pointing out the demarcation issue in the ID context but arguing that ID is religious and that teaching it would violate the establishment clause).
14. KUHN, *supra* note 4.
15. KUHN, *supra* note 4, at 153–54, 167–69, 177–78, 205–07.
16. *See* LARRY LAUDAN, SCIENCE AND RELATIVISM (1990) (dialogue between positivist, realist, pragmatist, and relativist, where each recognizes that relativist arguments on epistemology can be differentiated from the normative practice of a given scientific community).
17. *Id.*
18. *See generally* KUHN, *supra* note 4 (Kuhn repeatedly draws lines between science and philosophy or religion, and he discusses successful scientific revolutions as occurring through the use of the tools and problems of normal science, resulting in new scientific paradigms, but new paradigms must build on or improve previous theories to gain acceptance).

19. *Id.*
20. Kitzmiller, 400 F.Supp.2d at 718–23, 735–45; PENNOCK, *supra* note 1.
21. KUHN, *supra* note 4.
22. *Id.* at 177–78.
23. *Id.* at 167–69, 177–78.
24. Kitzmiller, 400 F.Supp.2d at 718–23, 735–45; PENNOCK, *supra* note 1.
25. *Id.*
26. KUHN, *supra* note 4, at 150–58.
27. FRANK S. RAVITCH, MASTERS OF ILLUSION: THE SUPREME COURT AND THE RELIGION CLAUSES 13–36 (2007); Frank S. Ravitch, *A Funny Thing Happened on the Way to Neutrality: Broad Principles, Formalism, and the Establishment Clause*, 38 GA. L. REV. 489, 498–523 (2004).
28. KUHN, *supra* note 4, at 153–54, 167–69, 177–78, 205–07.
29. Kitzmiller, 400 F.Supp.2d at 718–23, 735–45; PENNOCK, *supra* note 1.
30. KUHN, *supra* note 4, at 94, 129–30, 145–47, 153–54, 167–69, 177–79, 206–07.
31. PENNOCK, *supra* note 1.
32. Kitzmiller, 400 F.Supp.2d at 735–45.
33. KUHN, *supra* note 4, at 94, 129–30, 145–47, 153–54, 167–69, 177–79, 206–07.
34. *Id.*
35. *Id.*
36. *Id.*
37. *Id.*
38. *Id.*
39. *E.g.*, visit the Academic Freedom page on the Discovery Institute, Center for Science and Culture, Web site: http://www.discovery.org/csc/freeSpeechEvolCampMain.php.
40. Kitzmiller, 400 F.Supp.2d at 735–45.
41. Kitzmiller, 400 F.Supp.2d at 718–19, 746.
42. Kitzmiller, 400 F.Supp.2d at 735–45; Bishop, 926 F.2d 1066.

43. *See infra* notes 65–86, and accompanying text.
44. *See* KUHN, *supra* note 4 (referring to and defining *normal science* as the currently dominant scientific paradigm(s) and practices).
45. LAUDAN, *supra* note 16.
46. *See id.* (this argument is repeatedly made by the relativist and reflected in comments made by the others in the dialogue).
47. KUHN, *supra* note 4, at 150–58.
48. *Id.*; *see also* LAUDAN, supra 16 (argument made by relativist in dialogue).
49. LAUDAN, *supra* note 16 (excellent example of this debate presented in the form of a dialogue engaged in by archetypes of four of the major philosophical positions in the philosophy of science).
50. I have argued in the past that attacks on the possibility of value neutrality do have merit. RAVITCH, *supra* note 27; Ravitch, *supra* note 27.
51. FRANCIS J. BECKWITH, LAW, DARWINISM, AND PUBLIC EDUCATION: THE ESTABLISHMENT CLAUSE AND THE CHALLENGE OF INTELLIGENT DESIGN (2003); Francis J. Beckwith, *Public Education, Religious Establishment, and the Challenge of Intelligent Design*, 17 NOTRE DAME J.L. ETHICS & PUB. POL'Y 461 (2003); DeWolf et al., *supra* note 13.
52. JOHNSON, *supra* note 12; Wedge Document, *supra* note 12.
53. KUHN, *supra* note 4; LAUDAN, *supra* note 16.
54. KUHN, *supra* note 4; Ravitch, MASTERS OF ILLUSION, *supra* note 27.
55. Kitzmiller, 400 F.Supp.2d at 736.
56. RICHARD DAWKINS, THE GOD DELUSION (2006).
57. *See* Chapters 1 and 7.
58. *See* Chapters 3 and 7.
59. *Id.*
60. *See supra* notes 33–38, and accompanying text.
61. *Id.*
62. *Id.*
63. Kitzmiller, 400 F.Supp.2d at 736.

64. Larry Laudan, *The Demise of the Demarcation Problem*, *in* PHYSICS, PHILOSOPHY, AND PSYCHOANALYSIS: ESSAYS IN HONOR OF ADOLF GRÜNBAUM 111 (R. S. Cohen & Larry Laudan eds., 1983) (detailed discussion of the demarcation problem); Wexler, *supra* note 13 (pointing out the demarcation issue in the ID context but arguing that ID is religious and teaching it would violate the establishment clause).
65. Laudan, *supra* note 64, at 112–20.
66. *Id.*
67. *Id.* at 118, 124–25.
68. *See generally id.* (suggesting that there is no way to do so but that the question was not one of great importance in the first place).
69. Robert T. Pennock, *Can't Philosophers Tell the Difference between Science and Religion?: Demarcation Revisited*, SYNTHESE (April 2009).
70. Laudan, *supra* note 64, at 115, 121.
71. *Id.* at 122–23.
72. POPPER, *supra* note 2.
73. KUHN, *supra* note 4, at 94, 129–30, 145–47, 153–54, 167–69, 177–79, 206–07.
74. *Id.*; Pennock, *supra* note 69.
75. DeWolf et al., *supra* note 13.
76. Laudan, *supra* note 64, at 124–25.
77. *Id.* at 125.
78. *Id.* at 124–25.
79. *Id.*
80. DeWolf et al., *supra* note 13.
81. KUHN, *supra* note 4, at 94, 129–30, 145–47, 153–54, 167–69, 177–79, 206–07; Pennock, *supra* note 69.
82. *See* Chapters 3 and 6.
83. Edwards, 482 U.S. 578; Kitzmiller, 400 F.Supp.2d 707.
84. Edwards, 482 U.S. 578; *id.* at 599 (Powell, J., concurring).
85. CHARLES ALAN TAYLOR, DEFINING SCIENCE: A RHETORIC OF DEMARCATION (1996).
86. Pennock, *supra* note 69.

3. The Law and Intelligent Design

1. Kitzmiller v. Dover Area School District, 400 F.Supp.2d 707 (E.D. Pa. 2005); Selman v. Cobb County School District, 449 F.3d 1320 (11th Cir. 2006).
2. Kitzmiller, 400 F. Supp.2d 707.
3. 400 F.Supp.2d 707 (E.D. Pa. 2005).
4. *Id.*
5. *Id.*
6. *Id.* at 711–23, 735–46.
7. The argument, which would be a relatively weak one, would be based on equal access concepts. *See infra* notes 216–40, and accompanying text.
8. *Cf.* Kitzmiller, 400 F.Supp.2d at 711–23 (pointing out that constitutional issues arise because ID is a religious theory, not just because it is bad science).
9. *Id.* at 719, 721, 735–36.
10. *Id.* at 724–25, 727–29, 737–38, 740, 743–44.
11. *Id.* at 718–23, 735–45.
12. *Id.* at 735–46.
13. *Id.*
14. 482 U.S. 578 (1987).
15. *Id.*
16. Kitzmiller, 400 F.Supp.2d at 718.
17. *Id.* at 718–19.
18. *Id.*
19. *Id.* at 749–50, 755, 762.
20. *Id.* at 753.
21. *Id.* at 753–56.
22. *Id.* at 748–53.
23. *Id.*
24. *Id.* at 748, 752.
25. *Id.* at 714–35.
26. *Id.* at 735–46.
27. *See* David K. DeWolf, John G. West, & Casey Luskin, *Intelligent Design Will Survive Kitzmiller v. Dover*, 68 MONT. L. REV. 7, 25–42

(2007); Discovery Institute: Center for Science and Culture Web site, http://www.discovery.org/csc (featuring articles and news stories relating to such a claim).

28. 393 U.S. 97 (1968).
29. 482 U.S. 578 (1987).
30. Kitzmiller, 400 F.Supp.2d at 717–37.
31. *See generally id.*
32. U.S. CONST. AMEND. I.
33. 303 U.S. 1 (1947).
34. *See* Mary Ann Glendon & Raul F. Yanes, *Structural Free Exercise*, 90 MICH. L. REV. 477, 481–84 (1991).
35. AKHIL REED AMAR, THE BILL OF RIGHTS (1998).
36. *See supra* note 34; *see also* Everson v. Bd. of Educ., 330 U.S. 1 (1947) (incorporating the establishment clause).
37. *See, e.g.*, John C. Jeffries Jr. & James E. Ryan, *A Political History of the Establishment Clause*, 100 MICH. L. REV. 279, 305 (2001) (stating that 29 states have adopted establishment clauses); Toby J. Heytens, *School Choice and State Constitutions*, 86 VA. L. REV. 117, 127 (2000) (estimating that 20 state courts will interpret their establishment clauses more strictly than the Supreme Court's interpretation of the establishment clause).
38. *See generally* FRANK S. RAVITCH, MASTERS OF ILLUSION: THE SUPREME COURT AND THE RELIGION CLAUSES 72–86 (2007) (discussing the concept of seperationism and the various degrees and justifications for it).
39. *See generally id.* at 87–105 (exploring the concept of accomodationism and how it can function as a principle of religion clause interpretation).
40. *See generally id.*; *see also* Frank S. Ravitch, *A Funny Thing Happened on the Way to Neutrality: Broad Principles, Formalism, and the Establishment Clause*, 38 GA. L. REV. 489 (2004).
41. *See, e.g.*, Santa Fe Indep. Sch. Dist. v. Doe, 530 U.S. 290 (2000) (discussing the constitutionality of a school's policy of student-led, student-initiated prayer before football games); Good News Club v. Milford Cent. Sch., 533 U.S. 98 (2001) (equal access to government facilities by religious groups).

42. *See* Frank S. Ravitch, *Playing the Proof Game: Intelligent Design and the Law*, 113 PENN. ST. L. REV. 841 (2009).
43. 482 U.S. 578 (1987).
44. *Id.* at 585–94.
45. *See, e.g.*, Edwards, 482 U.S. 578 at 585–94; Wallace v. Jaffree, 472 U.S. 38, 56–61 (1985).
46. *See* Lee v. Weisman, 505 U.S. 577 (1992).
47. *See* Santa Fe Indep. Sch. Dist. v. Doe, 530 U.S. 290 (2000).
48. Kitzmiller, 400 F.Supp.2d at 712–35, 746–66.
49. *See* Lee, 505 U.S. 577.
50. *See, e.g.*, Lamb's Chapel v. Ctr. Moriches Union Free Sch. Dist., 508 U.S. 384, 398 (1993) (Scalia, J., concurring); Comm. for Pub. Educ. & Religious Liberty v. Nyquist, 413 U.S. 756, 662 (1973) (White, J., dissenting).
51. *See* Santa Fe, 530 U.S. at 314–17.
52. *Id.*
53. Lemon v. Kurtzman, 403 U.S. 602, 612 (1971).
54. *Id.*
55. *See* Epperson v. Arkansas, 393 U.S. 97, 106–09 (1968).
56. *Id.*
57. Lemon, 403 U.S. at 613.
58. *Id.* at 616–22.
59. *Id.* at 622–24.
60. *Cf.* Agostini v. Felton, 521 U.S. 203, 232–34 (1997) (rolling the entanglement prong of *Lemon* into the effects prong and dramatically limiting the importance of the divisiveness factor), *with* Santa Fe Indep. Sch. Dist. v. Doe, 530 U.S. 290, 311 (2000) (reasoning that a student election mechanism for determining whether prayer should be delivered before football games encourages divisiveness along religious lines in a public school setting and is at odds with the establishment clause).
61. Lemon, 403 U.S. at 612–13.
62. This seems clear given the reasoning in *Lemon*. Lemon, 403 U.S. 602, and the use of that reasoning in some later cases.
63. *See* Marcia S. Alembik, *The Future of the Lemon Test: A Sweeter Alternative for Establishment Clause Analysis*, 40 GA. L. REV. 1171,

1179 (2006) (discussing modifications, attempted replacements, and inconsistent results of the *Lemon* test).

64. Agostini, 521 U.S. at 232–34.
65. *Id.* at 223–34.
66. *Id.*
67. *Id.*
68. 536 U.S. 639 (2002).
69. *Id.*
70. 530 U.S. 290 (2000).
71. *Id.* at 314–17.
72. Kitzmiller v. Dover Area Sch. Dist., 400 F.Supp.2d 707 (E.D. Pa. 2005).
73. 465 U.S. 668 (1984).
74. *See* Frank S. Ravitch, *Religious Objects as Legal Subjects*, 40 WAKE FOREST L. REV. 1011, 1060–66 (2005).
75. *See, e.g.*, Kitzmiller, 400 F.Supp.2d at 714–35; Edwards v. Aguillard, 482 U.S. 578, 585–94 (1987).
76. *Id.*
77. *See* Ravitch, *supra* note 74.
78. *See* Santa Fe Indep. Sch. Dist. v. Doe, 530 U.S. 290 (2000); McCreary County v. ACLU, 545 U.S. 844 (2005).
79. Lynch, 465 U.S. at 687–94 (O'Connor, J., concurring).
80. *Id.* at 692–94; *see also* Edwards v. Aguillard, 482 U.S. 578, 585–94 (1987).
81. Kitzmiller, 400 F. Supp.2d at 714–35.
82. *See* Wallace v. Jaffree, 472 U.S. 38, 56–61 (1985); County of Allegheny v. ACLU, 492 U.S. 573, 592–94 (1989).
83. *See* Santa Fe, 530 U.S. at 308; McCreary County, 545 U.S. at 861.
84. Capitol Square Review Bd. v. Pinette, *infra*. this chapter (O'Connor, J., concurring).
85. *See* Capitol Square Review Bd. v. Pinette, *infra*. this chapter (O'Connor, J., concurring); *id.* (Stevens, J., dissenting).
86. *Id.* at 773 (O'Connor, J., concurring).
87. 482 U.S. 578 (1987).
88. *See* Wallace v. Jaffree, 472 U.S. 38, 56–61 (1985).
89. Edwards, 482 U.S. at 585–94.

90. *Id.*
91. *Id.*
92. *Id.*
93. 400 F. Supp.2d 707 (E.D. Pa. 2005).
94. 530 U.S. 290 (2000).
95. 505 U.S. 577 (1992).
96. *Id.*
97. *Id.*; *cf.* County of Allegheny v. ACLU, 492 U.S. 573, 655–79 (1989) (Kennedy, J., concurring in the judgment in part and dissenting in part) (applying the coercion test to two recurring holiday displays located on public property).
98. *See, e.g.*, McCreary County v. ACLU, 545 U.S. 844 (2005); Van Orden v. Perry, 545 U.S. 677 (2005).
99. *See* Lee, 505 U.S. at 586–99; Santa Fe Indep. Sch. Dist. v. Doe, 530 U.S. 290, 317 (2000).
100. In fact, in *Lee* itself, the Court phrased the coercion test this way: "It is beyond dispute that, at a minimum, the Constitution guarantees that government may not coerce anyone to support or participate in religion or its exercise, or otherwise act in a way which establishes a [state] religion or religious faith." Lee, 505 U.S. at 587.
101. *Id.* at 586.
102. *Id.*
103. *Id.* at 593–94.
104. *Id.* at 595–98.
105. *See, e.g.*, Engel v. Vitale, 370 U.S. 421 (1962) (fact that student prayer was voluntary does not lessen the indirect coercive pressure on religious minorities to conform); Sch. Dist. of Abington Twp. v. Schempp, 374 U.S. 203 (1963) (student exemption from prayer exercise does not alleviate the obvious pressure on children to participate).
106. Lee, 505 U.S. at 581; Santa Fe, 530 U.S. at 295–98.
107. *See* Cynthia V. Ward, *Coercion and Choice under the Establishment Clause*, 39 U.C. DAVIS L. REV. 1621, 1638 (2006) (discussing which types of religious exercises outside school prayer are subject to the indirect coercion test).

108. Lee, 505 U.S. at 593–95.
109. See Engel, 370 U.S. at 430–32; Schempp, 374 U.S. at 291–92.
110. William P. Marshall, *"We Know It When We See It": The Supreme Court Establishment*, 59 S. CAL. L. REV. 495, 500 (1986).
111. RAVITCH, *supra* note 38, at 13–36.
112. *See id.* at 13–36; Ravitch, *supra* note 40.
113. *See* Zelman v. Simmons-Harris, 536 U.S. 639 (2002).
114. RAVITCH, *supra* note 38, at 13–36; Ravitch, *supra* note 40.
115. *Id.*
116. 330 U.S. 1 (1947).
117. *Id.* at 15–16 (citing Reynolds v. United States, 98 U.S. 145, 164 (1879)).
118. Everson, 330 U.S. at 15–16; *id.* at 19, 28 (Jackson, J., concurring); *id.* at 29, 31–32 (Rutledge, J., dissenting).
119. *Id.* at 16–18.
120. *Id.* at 44–58 (Rutledge, J., dissenting).
121. PHILIP HAMBURGER, SEPARATION OF CHURCH AND STATE (2002); Arlin M. Adams & Charles J. Emmerich, *A Heritage of Religious Liberty*, 137 U. PA. L. REV. 1559 (1989).
122. Steven K. Green, *Of (Un)equal Jurisprudential Pedigree, Rectifying the Imbalance between Neutrality and Separationism*, 43 B.C. L. REV. 1111, 1121–25 (2002).
123. HAMBURGER, *supra.* note 121; Adams & Emmerich, *supra* note 121.
124. Douglas Laycock, *"Nonpreferential" Aid to Religion: A False Claim about Original Intent*, 27 WM. & MARY L. REV. 875–77, 878–902, 913–19 (1986); Adams & Emmerich, *supra* note 121; Theodore Sky, *The Establishment Clause, the Congress, and the Schools: An Historical Perspective*, 52 VA. L. REV. 1395, 1401–03, 1464–65 (1966).
125. Green, *supra* note 122.
126. Cass R. Sunstein, *Five Thesis on Originalism*, 19 HARV. J. L. & PUB. POL'Y. 311, 313 (1996).
127. RAVITCH, *supra* note 38.
128. Lemon, 403 U.S. at 614.

129. Ira C. Lupu, *The Lingering Death of Separationism*, 62 GEO. WASH. L. REV. 230, 249 (1994); *cf.* Michael W. McConnell, *Accommodation of Religion*, 1985 SUP. CT. REV. 1, 13–14 ("Excluding religious institutions and individuals from government benefits to which they would be entitled under neutral and secular criteria, merely because they are religious, advances secularism, not liberty").
130. Frederick Mark Gedicks, *A Two-Track Theory of the Establishment Clause*, 43 B.C. L. REV. 1071 (2002).
131. McConnell, *supra* note 129 (exploring connection between accommodation and the broad concept of liberty).
132. *Id.* at 24–27.
133. *Id.*
134. *Id.* at 29–32; Waltz, 397 U.S. at 673.
135. Frank S. Ravitch, *Religious Objects as Legal Subjects*, 40 WAKE FOREST L. REV. 1011 (2005).
136. 465 U.S. 668 (1984) (plurality opinion).
137. Ravitch, *supra* note 135.
138. *See* Lynch, 465 U.S. at 680, 683.
139. *Id.* at 680–86; *cf.* Steven D. Smith, *Separation and the "Secular": Reconstructing the Disestablishment Decision*, 67 TEX. L. REV. 955, 1002–03 (1989) (critics of the decision misunderstood the majority's secularization position; that position did not result from a conclusion that the crèche lost its religious meaning because of its placement but rather the majority employed a broad notion of the secular and found that the religious symbol served a secular purpose in the context involved).
140. Lynch, 465 U.S. at 711–12 (Brennan, J., dissenting).
141. *Id.*
142. *See, e.g., id.* (permissible accommodations are based on, consistent with, and further religious liberty).
143. Alan E. Brownstein, *Interpreting the Religion Clauses in Terms of Liberty, Equality, and Free Speech Values – A Critical Analysis of "Neutrality Theory" and Charitable Choice*, 13 NOTRE DAME J.L. ETHICS & PUB. POL'Y 243 (1999); Christopher L. Eisgruber &

Lawrence G. Sager, *Equal Regard*, *in* LAW & RELIGION: A CRITICAL ANTHOLOGY (Stephen M. Feldman ed., 2000).

144. *See, e.g.*, Douglas Laycock, *Formal, Substantive, and Disaggregated Neutrality toward Religion*, 39 DEPAUL L. REV. 993, 994 (1990) ("We can agree on the principle of neutrality without having agreed on anything at all").
145. RAVITCH, *supra* note 38, at 59–71.
146. *Cf.* Laycock, *supra* note 144, at 996 ("equality is an insufficient concept"; "claims about equality, or neutrality, always require further specification: equality with respect to what classification, for what purpose, in what sense, and to what extent?").
147. The classic *Lemon* test, *see* Lemon, 403 U.S. 602, was modified in Agostini v. Felton, 521 U.S. 203, 222–23 (1997) (folding the entanglement prong of *Lemon* into the effects prong and changing the Court's earlier focus on divisiveness as an element of entanglement).
148. Lynch, 465 U.S. at 690 (O'Connor, J., concurring); Wallace v. Jaffree, 472 U.S. 38, 56 (1985).
149. Lee v. Weisman, 505 U.S. 577 (1992).
150. Zelman, 122 S. Ct. 2460.
151. Kitzmiller, 400 F.Supp. 707.
152. *Id.* at 712–14.
153. *Id.* at 714–35, 765.
154. *Id.* at 746–65.
155. *Id.* at 765–66.
156. *Id.* at 745–46, 765.
157. *Id.*
158. *Id.* at 735–43.
159. *Id.* at 745–46.
160. *Id.* at 762–63.
161. *Id.* at 765.
162. *Id.* at 735–46, 763–65.
163. *Id.* at 745–46, 763–64 (school board policy and actions violated effects inquiry under endorsement and *Lemon* tests).
164. *Id.*

165. *Id.* at 748–62.
166. *Id.* at 763–65.
167. *Id.*
168. *Id.* at 748–62.
169. *Id.*
170. *Id.* at 711–46.
171. *See generally id.*
172. *Id.* at 720–23, 735–45.
173. *See generally id.*
174. *Id.* at 746–63.
175. 449 F.3d 1320 (11th Cir. 2006).
176. Selman v. Cobb County Sch. Dist. (*Selman I*), 390 F.Supp.2d 1286, 1292 (N.D. Ga. 2005), *vacated*, 449 F.3d 1320 (11th Cir. 2006).
177. CHARLES DARWIN, THE ORIGIN OF SPECIES 107–10, 375–76 (Bantam Books 1999) (1859); KENNETH R. MILLER, FINDING DARWIN'S GOD 53–54 (1999).
178. *Selman I*, 390 F.Supp.2d at 1303.
179. *Id.* at 1307.
180. *Id.*
181. *Id.*
182. *Id.* at 1312.
183. *Id.* at 1307.
184. Selman, 449 F.3d at 1338.
185. *Id.*
186. *Id.*
187. *Id.*
188. Epperson v. Arkansas, 393 U.S. 97, 107 (1968) (citing Abington Sch. Dist. v. Schempp, 374 U.S. 203, 225 (1963)).
189. Margaret G. Kivelson et al., *Galileo Magnetometer Measurements: A Stronger Case for a Subsurface Ocean at Europa*, SCIENCE (NEW SERIES), Aug. 2000, at 1340–43.
190. ROBERT PENNOCK, TOWER OF BABEL: THE EVIDENCE AGAINST THE NEW CREATIONISM (1999).
191. Discovery Institute, Center for Science and Culture, *Discovery Institute's Science Education Policy*, June 17, 2008, http://www.discovery.org/a/3164.

192. Richard Dawkins, *Natural "Knowledge" and Natural "Design,"* Science: Evolution and Biology page, May 16, 2006, http://richarddawkins.net/article,129,Natural-Knowledge-and-Natural-Design,Richard-Dawkins.
193. *See* Edwards, 482 U.S. at 591–92; Kitzmiller, 400 F.Supp.2d at 735.
194. 400 F.Supp.2d at 708–09.
195. *Id.* at 724, 729, 731–35, 762–63.
196. *See* Edwards, 482 U.S. at 628.
197. *See generally* Kitzmiller, 400 F.Supp.2d at 737–38.
198. Discovery Institute, Center for the Renewal of Science and Culture, *The Wedge*, available at http://ncse.com/webfm_send/747.
199. 393 U.S. at 103.
200. Kitzmiller, 400 F.Supp.2d at 720.
201. *Id.* at 735.
202. *See generally id.* at 731.
203. Edwards, 482 U.S. at 528 (Powell, J., concurring); *see, e.g.*, Cary v. Board of Educ., 598 F.2d 535 (10th Cir. 1979) (holding that the school board had a right to prescribe curriculum and textbooks and the teacher had no right to change or supplement); LeVake v. Indep. Sch. Dist. #656, 625 N.W.2d 502 (Minn. Ct. App. 2001) (granting summary judgment in favor of a school district where the school reassigned a science teacher who explicitly refused to teach the required course curriculum and asserted the right to academic freedom).
204. *See* Chapter 6.
205. *See, e.g.*, Cary v. Board of Educ., 598 F.2d 535 (10th Cir. 1979) (holding that the school board had a right to prescribe curriculum and textbooks and the teacher had no right to change or supplement); LeVake v. Indep. Sch. Dist. #656, 625 N.W.2d 502 (Minn. Ct. App. 2001) (granting summary judgment in favor of a school district where the school reassigned a science teacher who explicitly refused to teach the required course curriculum and asserted the right to academic freedom).
206. *See, e.g.*, Edwards, 482 U.S. at 583 (1987); Peloza v. Capistrano Unified Sch. Dist., 37 F.3d 517, 522 (9th Cir. 1994).

207. *See, e.g.*, Civil Rights Act of 1964, Equal Employment Opportunities, 42 U.S.C.A. § 2000e-2 (West 2010) (making it an unlawful employment practice to discriminate against any individual because of such individual's religion).
208. *See, e.g.*, *id.* at § 2000e-5(g)(1) (stating that a court may enjoin a respondent from engaging in such unlawful employment practice and other affirmative action as appropriate).
209. *See* Chapters 6 and 7.
210. *See* Kitzmiller, 400 F.Supp.2d at 736–37; PENNOCK, *supra* note 190; Discovery Institute, Center for Science and Culture, Academic Freedom page, http://www.discovery.org/csc/freeSpeech EvolCampMain.php.
211. *See infra* notes 216–40, and accompanying text.
212. KARL POPPER, LOGIC OF SCIENTIFIC DISCOVERY (1959).
213. *Id.*
214. THOMAS KUHN, THE STRUCTURE OF SCIENTIFIC REVOLUTIONS (3rd ed. 1996).
215. Kitzmiller, 400 F.Supp. at 735–46.
216. Ravitch, *supra* note 42.
217. Hague v. CIO, 307 U.S. 496 (1939).
218. Lamb's Chapel v. Ctr. Moriches Union Free Sch. Dist., 508 U.S. 384, 392–93 (1993).
219. Good News Club v. Milford Central Sch., 533 U.S. 98, 106–07 (2001) (citing Rosenberger v. Rector, 515 U.S. 819, 829 (1995)).
220. *Id.* at 107.
221. *Id.*
222. Capitol Square, 515 U.S. at 761 (citing Perry Educ. Ass'n v. Perry Local Educators' Ass'n., 460 U.S. 37, 45 (1983)).
223. *See, e.g.*, Good News Club v. Milford Central Sch., 533 U.S. 98, 105 (2001); Capitol Sq. Review and Advisory Bd. v. Pinette, 515 U.S. 753, 761 (1995); Lamb's Chapel v. Ctr. Moriches Union Free Sch. Dist., 508 U.S. 384, 390 (1993).
224. *See, e.g.*, Good News Club, 533 U.S. at 102; Lamb's Chapel, 508 U.S. at 387; Widmar v. Vincent, 454 U.S. 263, 269–70 (1981).
225. Good News Club, 533 U.S. at 106–07.

226. Good News Club, 533 U.S. at 108 (the district, however, disputed the scope of the forum).
227. *Cf. id.* at 107–10 (finding viewpoint discrimination where a group was excluded because of the religious perspective from which it addressed a variety of questions).
228. *Id.*
229. Capitol Square, 515 U.S. at 761; *see also* Church on the Rock v. City of Albuquerque, 84 F.3d 1273, 1279 (10th Cir. 1996) ("Content-based restrictions are subject to strict scrutiny. Viewpoint-based restrictions receive even more critical judicial treatment").
230. *Id.*
231. Rosenberger, 515 U.S. at 829–30.
232. Good News Club, 533 U.S. at 112–13.
233. Bartnicki v. Vopper, 532 U.S. 514, 544 (2001) (Rehnquist, J., dissenting) (implying that discrimination based on viewpoint is subject to strict scrutiny); Church on the Rock, 84 F.3d at 1279 ("Content-based restrictions are subject to strict scrutiny. Viewpoint-based restrictions receive even more critical judicial treatment"). The fact that the Court in *Good News Club* refused to decide whether viewpoint discrimination might be justified in to prevent violations of the establishment clause in rare circumstances at least leaves the question open. Good News Club, 533 U.S. at 112–13.
234. *Id.* at 106–12.
235. This is certainly implicit in *Good News Club*. *Id.* at 112–20.
236. *Id.*
237. Ravitch, *supra* note 42.
238. Good News Club v. Milford Central Sch., 533 U.S. 98, 105 (2001).
239. Hazelwood v. Kuhlmeier, 484 U.S. 260, 273 (1988).
240. *See* Chapters 1 and 6.

4. Theistic Evolution

1. Kenneth R. Miller, Finding Darwin's God (2002).
2. *Id.*

3. *The Proceedings of the Plenary Session on Scientific Insights into Evolution of the Universe and of Life*, PONTIFICIAE ACADEMIAE SCIENTIARVM ACTA (Werner Arber, Nicola Cabibbo, & Marcelo Sanchez Sorondo eds., 2008); Rabbi Jeremy Kalmanosky, *On That Yes, the Future of the World Depends*, http://www.uscj.org/cgi-bin/print.pl?On_That_Yes_The_Futu7488.html; Union for Reform Judaism Commission on Social Action of Reform Judaism to the 68th Union for Reform Judaism General Assembly, *Politicization of Science in the United States*, http://urj.org//about/union/governance/reso//?syspage=article&item_id=1943.
4. *See* MILLER, *supra* note 1 (explaining the widespread view that religion and evolution need not conflict).
5. *See* Chapter 1.
6. ALEXANDER HEIDEL, THE GILGAMESH EPIC AND OLD TESTAMENT PARALLELS (1963).
7. *Id.*
8. *Id.*
9. *Id.*
10. MILLER, *supra* note 1, at 238–39, 289–91.
11. *Id.* at 76–80.
12. *Id.*; *cf.* RONALD L. NUMBERS, THE CREATIONISTS: FROM SCIENTIFIC CREATIONISM TO INTELLIGENT DESIGN (exp. ed. 2006) (discussing in great detail the history of "flood geology" and its rejection by modern science).
13. HANS GEORG GADAMER, TRUTH AND METHOD 265–71 (Joel Weinsheimer & Donald G. Marshall trans., 2nd rev. ed. 1999).
14. *Id.*
15. *Id.* at 302–07.
16. *Id.* at 265–77, 302–07, 370, 374–75.
17. *Id.*
18. Frank S. Ravitch, Essay, *Scriptual Interpretation / Constitutional Interpretation*, 2009 MICH. ST. L. REV. 377 (2009).
19. NUMBERS, *supra* note 12.
20. GADAMER, *supra* note 13.

21. Discovery Institute, Wedge Document, 1998, available at http://www.antievolution.org/features/wedge.html; BARBARA FORREST & PAUL R. GROSS, CREATIONISM'S TROJAN HORSE (2004).
22. Kitzmiller v. Dover Area School District, 400 F.Supp.2d 707 (E.D. Pa. 2005); ROBERT PENNOCK, TOWER OF BABEL: THE EVIDENCE AGAINST THE NEW CREATIONISM (1999).
23. *See supra* notes 21–22, and accompanying text; PHILLIP JOHNSON, DARWIN ON TRIAL (1991).
24. RICHARD DAWKINS, THE GOD DELUSION (2006); DANIEL C. DENNETT, DARWIN'S DANGEROUS IDEA (1995).
25. *Id.*
26. FORREST & GROSS, *supra* note 21; PENNOCK, *supra* note 22; NUMBERS, *supra* note 12; JOHNSON, *supra* note 23.
27. JOHNSON, *supra* note 23; Wedge Document, *supra* note 21.
28. MILLER, *supra* note 1; DAWKINS, *supra* note 24; DENNETT, *supra* note 24; NUMBERS, *supra* note 12.
29. MILLER, *supra* note 1.
30. *Id.* at 102–03.
31. PENNOCK, *supra* note 22.
32. *Proceedings*, *supra* note 3; Kalmanosky, *supra* note 3; Union for Reform Judaism Commission, *supra* note 3; National Council of Churches, *Science, Religion, and Teaching of Evolution in Public School Science Classes*, http://www.ncccusa.org/pdfs/evolutionbrochurefinal.pdf; Episcopal Church, *Catechism of Creation Part II: Creation and Science*, http://episcopalchurch.org/19021_58398_ENG_HTM.htm; United Methodist Church, *Science and Technology*, http://www.umc.org/site/apps/nlnet/content3.aspx?c+1wL4KnN1LtH&b=30829&ct=6715227; Father George V. Coyne, *Science Does Not Need God. Or Does It? A Catholic Scientist Looks at Evolution*, CATHOLIC ONLINE, http://www.catholic.org/national/national_Story.php?id+18504; Denis R. Alexander, *Can a Christian Believe in Evolution?*, ALL TOGETHER, http://www.eauk.org/resources/idea/bigquestion/archive/2005/bq7.cfm.
33. *See, e.g.*, MILLER, *supra* note 1 (recounting examples of both).

34. *See, e.g.*, Peter Baker & Peter Slevin, *Bush Remarks on "Intelligent Design" Theory Fuel Debate*, WASHINGTON POST (Aug. 2, 2005) (noting that remarks by then president George W. Bush supporting intelligent design were supported by Christian conservatives but rejected by mainstream scientific organizations).
35. *See, e.g.*, David Klinghoffer, *Where Theistic Evolution Leads*, DISCOVERY INSTITUTE, http://www.discovery.org/a/10871 (May 19, 2009) (discussion of Kilnghoffer's entry on beliefnet Web site and his rejection of a somewhat caricatured view of theistic evolution).
36. MILLER, *supra* note 1.
37. *Id.* at 267–68.
38. *See supra* note 32.
39. National Council of Churches, *supra* note 32.
40. Kalmanosky, *supra* note 3.
41. United Methodist Church, *supra* note 32.
42. The text of this speech can be found online at http://www.ewtn.com/library/PAPALDOC/JP961022.HTM.
43. MILLER, *supra* note 1; *supra* note 32, and accompanying text.
44. DAWKINS, *supra* note 24; DENNETT, *supra* note 24.

5. Dawkins's Dilemma

1. KENNETH R. MILLER, FINDING DARWIN'S GOD: A SCIENTIST'S SEARCH FOR COMMON GROUND BETWEEN GOD AND EVOLUTION 165–91 (2002).
2. RICHARD DAWKINS, THE GOD DELUSION (2006).
3. *E.g.*, Dawkins asks whether science can demonstrate that there is a low or high probability of God, and he answers that science can do this. *Id.* at 137–51.
4. *See generally id.*
5. *Cf.* JOHN F. HAUGHT, GOD AND THE NEW ATHEISM: A CRITICAL RESPONSE TO DAWKINS, HARRIS, AND HITCHENS (2007) (attacking arguments made by the new atheists, including Dawkins, and pointing out that they are out to persuade people and have not defined what they mean by knowledge when they suggest knowing that a deity cannot be real).

6. DAWKINS, *supra* note 2.
7. *Id.* at 70–77, 137–88.
8. *See generally id.*
9. *Id.* at 51–94, 137–43, 190–240, 317–87.
10. *Id.*
11. *See* Chapter 2; MILLER, *supra* note 1.
12. *Id.* at 139–51, 185–86.
13. *Id.*
14. *Id.* at 51–94, 137–51, 190–240, 317–87.
15. *See generally id. See also* RICHARD DAWKINS, CLIMBING MOUNT IMPROBABLE (1997).
16. CHARLES DARWIN, THE ORIGIN OF SPECIES BY MEANS OF NATURAL SELECTION (Bantam Books 1999).
17. DAWKINS, *supra* note 2; DAWKINS, *supra* note 15; MILLER, *supra* note 1; EDWARD O. WILSON, FROM SO SIMPLE A BEGINNING: DARWIN'S FOUR GREAT BOOKS (2005).
18. MILLER, *supra* note 1; DAWKINS, *supra* note 15.
19. DAWKINS, *supra* note 2.
20. *See* Chapter 4; MILLER, *supra* note 1; *see also* HANS GEORG GADAMER, TRUTH AND METHOD 265–77, 302–07, 370, 374–75 (Joel Weinsheimer & Donald G. Marshall trans., 2nd rev. ed. 1999).
21. *Id.*
22. *Id.*
23. GADAMER, *supra* note 20, 265–77, 302–07, 370, 374–75.
24. *Id.*
25. *See* Chapter 4; MILLER, *supra* note 1.
26. DAWKINS, *supra* note 2.
27. *Cf. id.* at 70–77.
28. *Id.* at 97–99.
29. DAWKINS, *supra* note 2, at 190–240; *see also* CHRISTOPHER HITCHENS, GOD IS NOT GREAT: HOW RELIGION POISONS EVERYTHING (2007).
30. DAWKINS, *supra* note 2, at 190–240.
31. *Id.* at 97–99.
32. BRIAN GREENE, THE ELEGANT UNIVERSE: SUPERSTRINGS, HIDDEN DIMENSIONS, AND THE QUEST FOR THE ULTIMATE THEORY (1999).

33. *Id.*
34. *Id.*
35. DAWKINS, *supra* note 2, at 70–77, 137–88.
36. *See generally id.*
37. *See* Chapter 4.
38. *Id.*
39. DAWKINS, *supra* note 2.
40. *Id.* at 70–77, 137–88.
41. *See supra* notes 20–23, and accompanying text.
42. *See* Chapters 1 and 2.
43. *Id.*
44. ROBERT PENNOCK, TOWER OF BABEL: THE EVIDENCE AGAINST THE NEW CREATIONISM (1999).
45. *Id.*; MILLER, *supra* note 1.
46. *Id.*
47. *See* Chapters 1 and 2.
48. *Id.*
49. *Id.*
50. DAWKINS, *supra* note 2.
51. MILLER, *supra* note 1; DAWKINS, *supra* note 2; DAWKINS, *supra* note 15; KARL POPPER, LOGIC OF SCIENTIFIC DISCOVERY (1959).
52. *See* Chapters 1–3; Frank S. Ravitch, *Playing the Proof Game: Intelligent Design and the Law*, 113 PENN. ST. L. REV. 841 (2009) (discussing the proof game played by ID advocates).
53. DAWKINS, *supra* note 1, at 70–77, 137–88.
54. *See supra* notes 29–36, and accompanying text.
55. This is reflected throughout THE GOD DELUSION.
56. MILLER, *supra* note 1.
57. *See* Chapter 4.
58. *Id.*
59. MILLER, *supra* note 1.
60. DAWKINS, *supra* note 2.
61. RONALD L. NUMBERS, THE CREATIONISTS: FROM SCIENTIFIC CREATIONISM TO INTELLIGENT DESIGN (exp. ed. 2006).
62. MILLER, *supra* note 1.

63. DAWKINS, *supra* note 2; POPPER, *supra* note 51.
64. NUMBERS, *supra* note 61; PENNOCK, *supra* note 44.
65. MILLER, *supra* note 1, at 48–56.
66. *See* Chapter 4.
67. *Id.*; MILLER, *supra* note 1; DAWKINS, *supra* note 2; POPPER, *supra* note 51.
68. *See* Chapters 1–4.
69. DAWKINS, *supra* note 2.
70. *See generally* DAWKINS, *supra* note 2; *see also* DAWKINS, *supra* note 15.
71. *Id.*
72. *Id.* at 70–77.
73. *Id.*; POPPER, *supra* note 51.
74. *See* Chapters 2–4.
75. *See* Chapters 1 and 3.
76. Lisa Miller, *Darwin's Rottweiler*, NEWSWEEK (Oct. 5, 2009).
77. *See* Chapters 1–3.
78. *See* Chapter 1.
79. *Id.*
80. DAWKINS, *supra* note 2; DAWKINS, *supra* note 15; RICHARD DAWKINS, THE BLIND WATCHMAKER: WHY THE EVIDENCE OF EVOLUTION REVEALS A UNIVERSE WITHOUT DESIGN (1996).
81. *Cf.* MILLER, *supra* note 1, at 165–91 (discussing how militant antireligious stances by scientists may strengthen the position of creationists and ID advocates and hurt scientific and theistic evolutionary approaches).
82. *Id.*
83. DAWKINS, *supra* note 2; DAWKINS, *supra* note 80.
84. *Id.*
85. *See* Chapter 4; MILLER, *supra* note 1.

6. No Intelligence Allowed or No Intelligible Science?

1. EXPELLED: NO INTELLIGENCE ALLOWED (Premise Media Corp. 2008).
2. Peter Baker & Peter Slevin, *Bush Remarks on "Intelligent Design" Theory Fuel Debate*, WASHINGTON POST (Aug. 2, 2005)

(discussing a public statement by then president George W. Bush supporting ID).

3. RICHARD DAWKINS, THE GOD DELUSION (2006).
4. Kitzmiller v. Dover Area School District, 400 F.Supp.2d 707, 736 (E.D. Pa. 2005).
5. *See* Chapter 2.
6. International Society for Complexity, Information, and Design Center Web site, http://www.iscid.org/fellows.php, lists the following public university professors as fellows: Walter Bradley (mechanical engineering), Texas A&M University; J. Budziszewski (philosophy and political theory), University of Texas, Austin; John Angus Campbell (communications), University of Memphis; Russell W. Carlson (molecular biology), University of Georgia, Athens; Kenneth de Jong (linguistics), Indiana University, Bloomington; Daniel Dix (mathematics), University of South Carolina; Fred Field (linguistics), California State University; Guillermo Gonzalez (astronomy), Iowa State University; James Keener (mathematics and bioengineering), University of Utah; Robert C. Koons (philosophy), University of Texas, Austin; Stan Lennard (medicine), University of Washington; Scott Minnich (microbiology), University of Idaho; Martin Poenie (molecular cell and developmental biology), University of Texas, Austin; Carlos E. Puente (hydrology and theoretical dynamics), University of California, Davis; Henry F. Schaefer (quantum chemistry), University of Georgia, Athens; Jeffrey M. Schwartz (psychiatry/neuroscience), CLA Department of Psychiatry; Philip Skell (chemistry), Penn State University; Frederick Skiff (physics), University of Iowa; Karl D. Stephan (electrical engineering), Southwest Texas State University. Several of these names also appear on Discovery Institute's Center for Scientific Study Web site. It is possible that some of the people on the ISCIDC Web site did not realize that the organization is heavily focused on ID given the savvy marketing and description of the organization's purpose on the Web site, but when one looks closely at the organization, it is hard to miss the central focus on ID and related issues. The

preceding list does not include every public university faculty member who supports ID.

7. *E.g.*, the Discovery Institute's Center for Scientific Study lists the following as affiliates of the center and supporters of ID theory: Scott Minnich (microbiologist), University of Ohio; Dean Kenyon (emeritus biologist), San Francisco State University; and Henry Schaefer (chemist), University of Georgia. A number of these folks are listed on the ISCIDC list, see *supra* note 6, and that list includes other biologists and chemists. Neither list includes everyone who could be listed. The lists are meant to be illustrative of the fact that there are a number of faculty members at public universities whom ID theorists list as supporting the theory or aspects of it.
8. *See infra* Section II.A (providing detailed discussion of this issue).
9. *Id.*
10. *See infra* Section II.B (providing detailed discussion of this issue).
11. *Id.*
12. *See, e.g.*, THOMAS S. KUHN, THE STRUCTURE OF SCIENTIFIC REVOLUTIONS (3rd ed. 1996) (discussing paradigms in the sciences).
13. These questions are addressed *Supra* at Chapter 2.
14. *Id.*
15. *See* Discovery Institute, Center for Science and Culture, http://www.discovery.org/csc; International Society for Complexity, Information, and Design Center, http://www.iscid.org.
16. Kitzmiller, 400 F.Supp.2d at 718–22.
17. *See generally id.*
18. *See infra* notes 19–37, 185–202.
19. *Id.* at 719, 721, 735–36.
20. *Id.* at 724–25, 727–29, 737–38, 740, 743–44.
21. *Id.* at 718–23, 735–45.
22. *Id.* at 735–46.
23. 482 U.S. 578 (1987).
24. *Id.*
25. Kitzmiller, 400 F.Supp.2d at 718.
26. *Id.* at 721.

27. *Id.* at 720, 737.
28. *Cf. Id.* at 720–22, 737.
29. *Id.* at 720.
30. Bishop v. Aronov, 926 F.2d 1066 (11th Cir. 1991).
31. *Id.*; *see also infra this Chapter.*
32. KUHN, *supra* note 12 (discussing paradigms in the sciences and asserting that there is no superparadigm to decide between conflicting paradigms).
33. KUHN, *supra* note 12, at 153–54, 167–69, 177–78, 205–07.
34. *See generally* KUHN, *supra* note 12 (Kuhn repeatedly draws lines between science and philosophy or religion, and he discusses successful scientific revolutions as occurring through the use of the tools and problems of "normal science," resulting in new scientific paradigms, but new paradigms must build on or improve previous theories to gain acceptance).
35. *Id.*
36. KUHN, *supra* note 12, at 150–58.
37. KUHN, *supra* note 12, at 153–54, 167–69, 177–78, 205–07.
38. Kitzmiller, 400 F.Supp.2d at 718–23, 735–45; ROBERT PENNOCK, TOWER OF BABEL: THE EVIDENCE AGAINST THE NEW CREATIONISM (1999).
39. *See infra* this Chapter; *supra* Chapter 2.
40. *See supra* Chapter 2.
41. BRIAN GREENE, THE ELEGANT UNIVERSE: SUPERSTRINGS, HIDDEN DIMENSIONS, AND THE QUEST FOR THE ULTIMATE THEORY (1999).
42. Keary Davidson, *"Theory of Everything" Tying Researchers Up in Knots*, SAN FRANCISCO CHRONICLE (March 14, 2005) at A-6.
43. GREENE, *supra* note 41 at 222, 383–84.
44. For reasonably up-to-date information on the Large Hadron Collider, one can visit the CERN Web site at http://public.web.cern.ch/public/en/LHC/LHC-en.html.
45. For reasonably up-to-date information on the Fermi National Accelerator Laboratory, one can visit its Web site at http://www.fnal.gov.
46. Kitzmiller, 400 F.Supp.2d at 718–22, 735–45.

47. *Id.*; PENNOCK, *supra* note 38.
48. *Id.*
49. *Id.*
50. *Id.*
51. *Id.*
52. *Id.*
53. *Id.*
54. GREENE, *supra* note 41; Davidson, *supra* note 42.
55. GREENE, *supra* note 41.
56. *Id.*
57. *Id.*; Davidson, *supra* note 42.
58. GREENE, *supra* note 41.
59. *Cf.* KUHN, *supra* note 12, at 94, 129–30, 145–47, 153–54, 167–69, 177–79, 206–07 (suggesting that without some persuasive evidence that can be accepted by the community of scientists, new scientific paradigms are unlikely to be persuasive or lead to a scientific revolution).
60. *Id.*
61. GREENE, *supra* note 41, at 137–40.
62. *Id.* at 139–40.
63. *See generally* GREENE, *supra* note 41 (explaining the basics of superstring theory).
64. *Id.* at 283–306.
65. *Id.* at 201–05.
66. *Id.* at 184–227.
67. *Id.* at 308–19.
68. *Id.* at 203–04.
69. *Id.* at 308–19.
70. *Id.* at 383–87.
71. LEE SMOLIN, THE TROUBLE WITH PHYSICS: THE RISE OF STRING THEORY, THE FALL OF SCIENCE, AND WHAT COMES NEXT (Houghton Mifflin 2006); KARL R. POPPER, CONJECTURES AND REFLECTIONS: THE GROWTH OF SCIENTIFIC KNOWLEDGE (5th ed. 2002) (classic text addressing Popper's theories on science).
72. *Cf. generally Id.* (Noting the meticulous path of string theory calculations and the importance of the fact that they ultimately

worked out); *Id.* at 308–319 (discussing importance of Witten's M-Theory in uniting previously conflicting versions of string theory).

73. *Id.* at 20, 383–87.
74. *Id.* at 222, 383–87
75. Smolin, *supra* note 103.
76. Greene, *supra* note 73 at 222, 383–87.
77. *Id.* at 20.
78. *Id.*
79. *See, e.g.*, Edwards v. California University of Pennsylvania, 156 F.3d 488, 491 (3rd Cir. 1998) ("[A] public university professor does not have a First Amendment right to decide what will be taught in the classroom"); Keen v. Penson, 970 F.2d 252, 257 (7th Cir. 1992) ("This Court has recognized the supremacy of the academic institution in matters of curriculum content"); Bishop v. Aronov, 926 F.2d 1066, 1075 (11th Cir. 1991) (noting that university officials may control curricular decisions); Scallet v. Rosenblum, 911 F.Supp. 999, 1011 (W.D. Va. 1996) (recognizing that the schools have a right to determine their own curricula, which must be followed); Cooper v. Ross, 472 F. Supp. 802, 809 (E.D. Ark. 1979) ("[A] state university has the undoubted right to prescribe its curriculum, to select its faculty and students and evaluate their performances, and to define and maintain its standards of academic accomplishment").
80. *See, e.g.*, Keyshian v. Board of Regents of University of the State of New York, 385 U.S. 589, 603 (1967) ("Our Nation is deeply committed to safeguarding academic freedom, which is of transcendent value to all of us and not merely to the teachers concerned. That freedom is therefore a special concern of the First Amendment, which does not tolerate laws that cast a pall of orthodoxy over the classroom"); Scallet v. Rosenblum, 911 F.Supp. 999, 1014 (noting the importance of academic freedom); Hardy v. Jefferson Community College, 260 F.3d 671, 679 (6th Cir. 2001) (involving a professor using profane language; however, because the course was one dealing with interpersonal communication, the court found it to be within the ambit of the curriculum, despite the university's

protests); Vanderhurst v. Colorado Mountain College Dist., 208 F.3d 908, 913 (10th Cir. 2000) (noting that academic freedom is a "special concern" of the First Amendment); Bishop v. Aronov, 926 F.2d 1066, 1075 (11th Cir. 1991) (noting that there is a strong recognition of academic freedom as it relates to the First Amendment).

81. *See, e.g.*, Edwards v. California University of Pennsylvania, 156 F.3d 488, 491 (3rd Cir. 1998) (noting that while a professor may advocate for a change in the curriculum *outside* the classroom, the professor may not use those materials *in* the classroom); Scallet v. Rosenblum, 911 F.Supp. 999, 1014 (noting that not all speech will implicate the First Amendment and academic freedom); Keen v. Penson, 970 F.2d 252, 257 (7th Cir. 1992) (recognizing that there may be conflicts between academic freedom and control over the curriculum that require some limiting of academic freedom); Cohen v. San Bernardino Valley College, 92 F.3d 968, 972 (9th Cir. 1996) (acknowledging the potential limitations on academic freedom); Bishop v. Aronov, 926 F.2d 1066, 1075 (11th Cir. 1991) (recognizing that the university's interest in having its courses taught without religious bias outweighed the countervailing concerns related to academic freedom within the curriculum); Clark v. Holmes, 474 F.2d 928, 931 (7th Cir. 1973) ("We do not conceive academic freedom to be a license for uncontrolled expression at variance with established curricular contents and internally destructive of the proper functioning of the institution. First Amendment rights must be applied in light of the special characteristics of the environment in the particular case").

82. Bishop, 926 F.2d 1066; *see also* Edwards v. California University of Pennsylvania, 156 F.3d 488, 491 (3rd Cir. 1998) (involving courses on educational media, unrelated to religion, being taught with a religious bias). Other cases related to academic freedom and curricula deal with secular concerns. For example, one prominent case dealt with, among other things, a professor's in-class discussions related to diversity in a first-year required writing course. Scallet v. Rosenblum, 911 F.Supp. 999, 1003 (W.D. Va. 1996). The court found that despite the fact that the issue was one of public

concern, the university's interest in a consistent curriculum outweighed the professor's First Amendment rights and was not protected. *Id.* at 1017.

83. *See, e.g.*, Vanderhurst v. Colorado Mountain College Dist., 208 F.3d 908, 914 (10th Cir. 2000) ("[W]hether . . . [the] termination reasonably related to the College's legitimate pedagogical interests is the test for determining whether his speech fell within the ambit of First Amendment protection"); Scallet v. Rosenblum, 911 F.Supp. 999, 1016 (W.D. Va. 1996) (noting that a professor's use of certain materials violated the university's legitimate pedagogical interests; however, the case also notes that the pedagogical concerns are less forceful at the university level than at lower educational levels such as high school); Silva v. University of New Hampshire, 888 F.Supp. 293, 314 (D.N.H. 1994) (recognizing a right to protect valid pedagogical purposes but finding the policy in this case too subjective to merit protection).

84. *See, e.g.*, Bonnell v. Lorenzo, 241 F.3d 800, 803 (6th Cir. 2000) (involving use of profanity during class by a professor and finding that the interests of the professor were outweighed by the university's concerns); Bishop v. Aronov, 926 F.2d 1066, 1075 (11th Cir. 1991) (recognizing that the university's interest in having its courts taught without religious bias outweighed the countervailing concerns related to academic freedom within the curriculum); Scallet v. Rosenblum, 911 F.Supp. 999, 1016–17 (W.D. Va. 1996) (recognizing a balancing test, and in this case, the professor's interest was outweighed by the university's interest in having its curriculum taught without significant disruption). *But see, e.g.*, Hardy v. Jefferson Community College, 260 F.3d 671, 679 (6th Cir. 2001) (discussing a situation where the pedagogical interests did not outweigh the activities and speech of the professor); Cohen v. San Bernardino Valley College, 92 F.3d 968, 972 (9th Cir. 1996) (while the lower court found that, on the balance, the professor's interest in teaching the controversial material was outweighed by the university's interest in effective education as determined by its curriculum, the appellate court found that the university's policies in this regard were too vague to be enforceable).

85. *See, e.g.*, Edwards v. California University of Pennsylvania, 156 F.3d 488, 491 (3rd Cir. 1998) (finding it unnecessary to inquire further into the issue of the First Amendment standard given at the trial level because "a public university professor does not have a *First Amendment* right to decided what will be taught in the classroom"); Martin v. Parrish, 805 F.2d 583, 583 (5th Cir. 1986) (determining that the First Amendment concerns related to academic freedom did not apply as the language in question was unrelated to the subject matter of the class); Hetrick v. Martin, 480 F.2d 705, 708 (6th Cir. 1973) (determining that a university may dismiss a professor based on disagreements with the professor's "pedagogical attitudes").
86. *See, e.g.*, Zelman v. Simmons-Harris, 536 U.S. 639 (2002) (formal neutrality); Good News Club v. Milford Central School, 533 U.S. 98 (2001) (equal access).
87. *See supra* Chapter 3.
88. *See, e.g.*, Hardy v. Jefferson Community College, 260 F.3d 671, 679 (6th Cir. 2001) (involving the use of profane language in the classroom); Vanderhurst v. Colorado Mountain College Dist., 208 F.3d 908, 911 (10th Cir. 2000) (involving a series of profane and offensive remarks unrelated to the curriculum); Edwards v. California University of Pennsylvania, 156 F.3d 488, 491 (3rd Cir. 1998) (involving courses being taught with a religious bias); Cohen v. San Bernardino Valley College, 92 F.3d 968, 972 (9th Cir. 1996) (involving intentionally shocking discussions regarding profane language and controversial topics, including cannibalism and consensual sex with children); Bishop v. Aronov, 926 F.2d 1066, 1075 (11th Cir. 1991) (involving a university's concern that courses not be taught with a religious bias).
89. *See, e.g.*, Hardy v. Jefferson Community College, 260 F.3d 671, 679 (6th Cir. 2001) (involving profane language and using such terms as *nigger* and *bitch* during class discussions on social deconstructivism); Bonnell v. Lorenzo, 241 F.3d 800, 803 (6th Cir. 2000) (involving the use of profanity during class by a professor); Vanderhurst v. Colorado Mountain College Dist., 208 F.3d 908, 911 (10th Cir. 2000) (outlining a series of vulgar and offensive

remarks the professor made unrelated to the course material or, in many cases, any educational purpose whatsoever); Cohen v. San Bernardino Valley College, 92 F.3d 968, 972 (9th Cir. 1996) (involving profane language and controversial topics that were arguably outside the curriculum of the class); Bishop v. Aronov, 926 F.2d 1066, 1075 (11th Cir. 1991) (inserting religious material and perspectives into a course that did not deal with religion but instead with science); Martin v. Parrish, 805 F.2d 583, 583 (5th Cir. 1986) (involving profanity in the classroom).

90. *See, e.g.*, Scallet v. Rosenblum, 911 F.Supp. 999, 1011 (involving a required Analysis and Communication course and a professor inserting his opinions on matters outside the core course curriculum).
91. 926 F.2d 1066 (11th Cir. 1991).
92. *Id.* at 1068.
93. *Id.*
94. *Id.* at 1069.
95. *Id.* at 1068.
96. *Id.*
97. *Id.* at 1068, 1076–77.
98. *Id.* at 1076–77.
99. *Id.* at 1071.
100. *Id.* at 1074.
101. *Id.*
102. *Id.* at 1076.
103. *Id.*
104. *Id.* at 1076–77.
105. *Id.*
106. *Id.* at 1074, 1076–77.
107. *Cf. id.* at 1074 (using cases that relied on "legitimate pedagogical interests" language but not using that exact language as set forth in those cases).
108. *Id.* at 1074–75.
109. *Id.*
110. *Id.* at 1075 (emphasis added).
111. *Id.* at 1076.

112. *See* Section II.D.
113. On the basis of the case law, public universities have a fairly wide latitude to determine what will be taught. In this context, excluding ID from the science curriculum is in line with other curricular decisions. *Cf.* Edwards v. California University of Pennsylvania, 156 F.3d 488, 491 (3rd Cir. 1998) ("[A] public university professor does not have a First Amendment right to decide what will be taught in the classroom"); Keen v. Penson, 970 F.2d 252, 257 (7th Cir. 1992) ("This Court has recognized the supremacy of the academic institution in matters of curriculum content"); Bishop v. Aronov, 926 F.2d 1066, 1075 (11th Cir. 1991) (noting that university officials may control curricular decisions); Scallet v. Rosenblum, 911 F.Supp. 999, 1011 (W.D. Va. 1996) (recognizing that the schools have a right to determine their own curricula, which must be followed); Cooper v. Ross, 472 F. Supp. 802, 809 (E.D. Ark. 1979) ("[A] state university has the undoubted right to prescribe its curriculum, to select its faculty and students and evaluate their performances, and to define and maintain its standards of academic accomplishment"). The ability of a university to control its science curriculum appears to be especially true as it pertains to ID and science as at least one prominent decision has determined that ID is not science. Kitzmiller, 400 F. Supp.2d 707.
114. *See, e.g.*, Hardy v. Jefferson Community College, 260 F.3d 671, 678 (6th Cir. 2001) (reviewing a number of secondary school cases to support its determination); Vanderhurst v. Colorado Mountain College Dist., 208 F.3d 908, 912–13 (10th Cir. 2000) (same); Bishop v. Aronov, 926 F.2d 1066, 1072–74 (11th Cir. 1991) (same); Scallet v. Rosenblum, 911 F.Supp. 999, 1016–17 (W.D. Va. 1996) (same).
115. *See, e.g.*, Brown v. Arementi, 247 F.3d 69, 75 (3rd Cir. 2001) (discussing on what basis a university may regulate professors based on its own pedagogic concerns over academic freedom); Vanderhurst v. Colorado Mountain College Dist., 208 F.3d 908, 914 (10th Cir. 2000) ("[W]hether . . . [the] termination reasonably related to the College's legitimate pedagogical interests is the test for

determining whether his speech fell within the ambit of First Amendment protection"); Bishop v. Aronov, 926 F.2d 1066, 1071, 1074 (11th Cir. 1991) ("[E]ducators do not offend the First Amendment by exercising editorial control over the style and content of student [or professor] speech in school-sponsored expressive activities so long as their actions are reasonably related to legitimate pedagogical concerns"); Scallet v. Rosenblum, 911 F.Supp. 999, 1014 (W.D. Va. 1996) (determining that the speech in question was seen as university speech); Silva v. University of New Hampshire, 888 F.Supp. 293, 314 (D.N.H. 1994) (recognizing a right to protect valid pedagogical purposes but finding the policy in this case too subjective to merit protection).

116. *See, e.g.*, Edwards v. California University of Pennsylvania, 156 F.3d 488, 492 (3rd Cir. 1998) (noting that in this case, regulation of the speech was allowed because the university could be considered the speaker, through the professor, and could make decisions as to the content of its own derivative speech); Bishop v. Aronov, 926 F.2d 1066, 1071 (11th Cir. 1991) (recognizing that the university has an interest in the professor disseminating his beliefs under the guise of university instruction); Scallet v. Rosenblum, 911 F.Supp. 999, 1014 (W.D. Va. 1996) (holding that the speech is university speech).

117. *See, e.g.*, Edwards v. California University of Pennsylvania, 156 F.3d 488, 492 (3rd Cir. 1998) (discussing why a university may control a private individual's speech when it is done in a manner that makes it, in reality, university speech); Bishop v. Aronov, 926 F.2d 1066, 1073 (11th Cir. 1991) ("While a student's expression can be more readily identified as a thing independent of the school, a teacher's speech can be taken as directly and deliberately representative of the school. Hence, where the in-class speech of a teacher is concerned, the school has an interest not only in preventing interference with the day-to-day operation of its classrooms as in *Tinker*, but also in scrutinizing expressions that "the public might reasonably perceive to bear its imprimatur"); Scallet v. Rosenblum, 911 F.Supp. 999, 1014 (W.D. Va. 1996) (finding that the speech in question was public, not private).

But see Bonnell v. Lorenzo, 241 F.3d 800, 811 (6th Cir. 2000) (involving a case of mixed private and public speech).

118. *See, e.g.*, Vanderhurst v. Colorado Mountain College Dist., 208 F.3d 908, 913–14 (10th Cir. 2000) (focusing on the free speech rights of a government employee); Levin v. Harleston, 966 F.2d 85, 88 (2nd Cir. 1992) (focusing on the free speech rights of the professor in question).
119. *See, e.g.*, Brown v. Arementi, 247 F.3d 69, 75 (3rd Cir. 2001) (finding that the professor's contentions regarding the grading policy were not matters of public concern); Hardy v. Jefferson Community College, 260 F.3d 671, 679 (6th Cir. 2001) (focusing on aspects of the speech related to speaking on a matter of public concern); Dambrot v. Central Michigan University, 55 F.3d 1177, 1188 (6th Cir. 1995) (focusing on whether the speech was a matter of public concern); Blum v. Schlegel, 18 F.3d 1005, 1011 (2nd Cir. 1994) (involving, in part, whether a law school professor's advocating for legalized marijuana was a matter of public concern); Keen v. Penson, 970 F.2d 252, 258 (7th Cir. 1992) (discussing the balancing of comments on matters of public concern but failing to find that it was a matter of public concern in this particular case); Martin v. Parrish, 805 F.2d 583, 583 (5th Cir. 1986) (noting that the test for matters involving public employees is whether their speech touches on a matter of public concern).
120. *See, e.g.*, Bishop, 926 F.2d at 1072 (discussing both the free speech rights of government employees generally and the weighing of interests on matters of public concern); Bonnell v. Lorenzo, 241 F.3d 800, 803 (6th Cir. 2000) (primarily using a public concern approach but also discussing the rights of public employees); Marinoff v. City College of New York, 357 F. Supp.2d 672 (S.D.N.Y. 2005) (same); Scallet v. Rosenblum, 911 F.Supp. 999, 1014 (W.D. Va. 1996) (same).
121. *Cf.* Edwards v. California University of Pennsylvania, 156 F.3d 488, 491 (3rd Cir. 1998) ("[A] public university professor does not have a First Amendment right to decide what will be taught in the classroom"); Keen v. Penson, 970 F.2d 252, 257 (7th Cir. 1992) ("This Court has recognized the supremacy of the academic

institution in matters of curriculum content"); Bishop v. Aronov, 926 F.2d 1066, 1075 (11th Cir. 1991) (noting that university officials may control curricular decisions); Scallet v. Rosenblum, 911 F.Supp. 999, 1011 (W.D. Va. 1996) (recognizing that the schools have a right to determine their own curricula, which must be followed); Cooper v. Ross, 472 F. Supp. 802, 809 (E.D. Ark. 1979) ("[A] state university has the undoubted right to prescribe its curriculum, to select its faculty and students and evaluate their performances, and to define and maintain its standards of academic accomplishment").

122. Bishop, 926 F.2d at 1076.
123. Kitzmiller, 400 F.Supp.2d at 737–38, 744–45.
124. *See, e.g.*, Donna Euben & Barbara Lee, *Faculty Discipline: Legal and Policy Issues in Dealing with Faculty Misconduct*, 32 J.C. & U.L. 241, 251 (2006) (describing some of the various ways in which universities sanction tenured professors, including revocation, and the various standards); Michael Olivas, *Review of Steven G. Poskanzer's Higher Education Law: The Faculty*, 30 J.C. & U.L. 225 (providing an overview of some cases dealing with tenure revocation and the various ways it is dealt with). *Compare* Kansas State University Handbook, section C 31.5 (dealing with tenure revocation), http://www.k-state.edu/academicservices/fhbook/fhsecc.html#31.5, *with* University of New Mexico Faculty Handbook, section 6.2.2(B) (dealing with dismissal of tenured faculty), http://www.unm.edu/~handbook/newhb.html, *and* University of Georgia Policy for Review of Tenured Faculty, http://www.uga.edu/provost//polproc/revtefac.html, *and* Michigan Technological Institute Academic Tenure and Promotion, http://www.admin.mtu.edu/admin/boc/policy/ch16/ch16p4.htm. The preceding examples demonstrate just some of the wide-ranging variances in tenure policy.
125. *Cf.* STEVEN G. POSKANZER, HIGHER EDUCATION LAW: THE FACULTY 209–17 (2002) (explaining circumstances under which for-cause tenure revocation is generally done); *supra* note 124, and

accompanying text (referring to diversity of public university tenure revocation policies).

126. *See supra* note 124, and accompanying text.

127. It is likely that such a decision would fall under the ambit of decisions that involve a university's right to determine the curriculum and thus implicitly what activities it will provide funding for. *Cf.* Edwards v. California University of Pennsylvania, 156 F.3d 488, 491 (3rd Cir. 1998) ("[A] public university professor does not have a First Amendment right to decide what will be taught in the classroom"); Bishop v. Aronov, 926 F.2d 1066, 1075 (11th Cir. 1991) (noting that university officials may control curricular decisions); Scallet v. Rosenblum, 911 F.Supp. 999, 1011 (recognizing that the schools have a right to determine their own curricula, which must be followed); Keen v. Penson, 970 F.2d 252, 257 (7th Cir. 1992) ("This Court has recognized the supremacy of the academic institution in matters of curriculum content"); Cooper v. Ross, 472 F. Supp. 802, 809 (E.D. Ark. 1979) ("[A] state university has the undoubted right to prescribe its curriculum, to select its faculty and students and evaluate their performances, and to define and maintain its standards of academic accomplishment").

128. *Cf.* Urofsky v. Gilmore, 216 F.3d 401, 410 (4th Cir. 2000) (determining that a Virginia statute limiting access to sexually explicit material for research did not violate the academic freedom of the professors; this is similar to limiting, by not giving credit for, research related to ID); Bishop v. Aronov, 926 F.2d 1066, 1075 (11th Cir. 1991) (noting that university officials may control curricular decisions, likely including ones involving what research will be credited within the department); Scallet v. Rosenblum, 911 F.Supp. 999, 1011 (recognizing that the schools have a right to determine their own curricula, which must be followed); Keen v. Penson, 970 F.2d 252, 257 (7th Cir. 1992) ("This Court has recognized the supremacy of the academic institution in matters of curriculum content"); Cooper v. Ross, 472 F. Supp. 802, 809 (E.D. Ark. 1979) ("[A] state university has the undoubted right

to prescribe its curriculum, to select its faculty and students and evaluate their performances, and to define and maintain its standards of academic accomplishment"). *See also* Dow Chemical v. Allen, 672 F.2d 1262 (7th Cir. 1982) (the court determined that, related to academic freedom, "it is clear that whatever constitutional protection is afforded by the First Amendment extends as readily to the scholar in the laboratory as to the teacher in the classroom"; this proposition may be read to support the preceding cases in the sense that the professors have a wide latitude *within* their research area but cannot simply research outside subjects like ID (just as they cannot simply teach ID) without university approval of the curriculum and research).

129. Kitzmiller, 400 F.Supp.2d at 737–38, 744–45.
130. *Id.* at 735–46; *supra* Chapters 1 & 2.
131. POSKANZER, *supra* note 125 at 208–17.
132. *See supra* note 124, and accompanying text.
133. POSKANZER, *supra* note 125 at 209–11.
134. Euban & Lee, *supra* note 124 at 301–03.
135. POSKANZER, *supra* note 125 at 211–17.
136. *See supra* notes 132–34, and accompanying text.
137. *Id.*
138. *Compare* National Endowment for the Arts v. Finley, 524 U.S. 569 (1998) (government as speaker); Rust v. Sullivan, 500 U.S. 173 (1991) (same), *with* U.S. v. National Treasury Employees Union, 513 U.S. 454 (1995) (public employee speech); Rankin v. McPherson, 483 U.S. 378 (1987) (same); Connick v. Myers, 461 U.S. 138 (1983) (same); Pickering v. Bd. of Education, 391 U.S. 563 (1968) (same).
139. *Id.*
140. Bishop, 926 F.2d at 1074, 1076–77.
141. *Id.*
142. 400 F.Supp.2d 707.
143. *Id.* at 714–35.
144. *Id.* at 708–09.
145. Bishop, 926 F.2d at 1076–77.

146. Santa Fe Indep. School Dist. v. Doe, 530 U.S. 290 (2000); Edwards v. Aguillard, 482 U.S. 578 (1987); Wallace v. Jaffrree, 472 U.S. 38 (1985); Kitzmiller, 400 F.Supp.2d 707.
147. Santa Fe, 530 U.S. 290; Lee v. Weisman, 505 U.S. 577 (1992).
148. Santa Fe, 530 U.S. 290; Edwards, 482 U.S. 578; Kitzmiller, 400 F.Supp.2d 707.
149. The classroom involves a captive audience setting, Abington Township v. Schempp, 374 U.S. 203 (1963), and the professor at a public university is a state employee. Scholarship is generally engaged in by the professor, perhaps with help from research and lab assistants, who can choose to apply for a position. *Cf.* Bishop, 926 F.2d 1066.
150. Bishop, 926 F.2d at 1074, 1076–77 (discussing possibility of students feeling coerced).
151. *Id.*
152. *Cf.* Bishop, 926 F.2d at 1074, 1076–77 (discussing the possibility of students feeling coerced); Kitzmiller, 400 F.Supp.2d at 714–35 (endorsement in the high school setting).
153. Santa Fe, 530 U.S. 308–10; Edwards, 482 U.S. at 585–87; Lynch v. Donnelly, 465 U.S. 668, 688 (1994) (O'Connor, J., concurring).
154. *Cf.* Bishop, 926 F.2d 1076–77 (addressing the university's potential concern that Bishop's religious messages in the course could create an "appearance of proselytizing by a professor").
155. Kitzmiller, 400 F.Supp.2d 707.
156. *Id.*
157. *Cf. id.* at 714–45 (disclaimer regarding evolution and ID in high school science courses).
158. *Id.*; Edwards, 482 U.S. 578; *cf.* Bishop, 926 F.2d at 1076–77 (noting that this would be a valid concern for a university).
159. Bishop, 926 F.2d at 1074, 1076–77.
160. Kitzmiller, 400 F.Supp.2d 735–45.
161. *Id.* at 718–45.
162. *Id.* at 714–35.
163. *Cf. id.* at 714–45 (teaching or referring in a disclaimer to ID as valid scientific theory endorses religion in the high school context); Bishop, 926 F.2d at 1074, 1076–77 (endorsement or

promotion of religion is a valid concern when a professor uses a religious approach in teaching a secular science course).

164. Bishop, 926 F.2d at 1074, 1076–77.
165. Kitzmiller, 400 F.Supp.2d at 746–64.
166. *See, e.g.*, Edwards, 582 U.S. at 582–83 (violation of any one prong of the *Lemon* test is adequate to demonstrate an establishment clause violation under that test); ACLU of New Jersey v. Black Horse Pike Regional Bd. of Educ., 84 F.3d 1471, 1488 (3rd Cir. 1996) (en banc) (no need to address the entanglement prong of *Lemon* when other prongs are met).
167. *See* Sections II.A, II.B., and II.C.
168. *Id.*
169. For information on the Raelian cult, including the Raelians' argument for alien "intelligent design," *see* http://www.rael.org.
170. Susan Palmer, *The Raelian Movement International*, *in* NEW RELIGIONS AND THE NEW EUROPE 200 (Robert Towler ed., 1995).
171. *See supra* this Section.
172. Santa Fe, 530 U.S. at 301–07; Lee v. Weisman, 505 U.S. 577 (1992).
173. *Cf.* Bishop, 926 F.2d at 1068–69, 1074, 1076–77 (addressing potential coercion from a religious approach to a general science course).
174. Lee, 505 U.S. at 577.
175. *Cf.* Bishop, 926 F.2d at 1074, 1076–77 (noting potential coercion when a religious approach is used in a university science class).
176. *See id.* at 1068–69 (discussing student complaints regarding a religious approach in a science course).
177. DAWKINS, *supra* note 3.
178. *See* Chapters 1–5.
179. DAWKINS, *supra* note 3.
180. KENNETH R. MILLER, FINDING DARWIN'S GOD (2002); EDWARD O. WILSON, FROM SO SIMPLE A BEGINNING: DARWIN'S FOUR GREAT BOOKS (2005), *repr. in Can Biology Do Better Than Faith?*, NEW SCIENTIST (Nov. 2005).
181. MICHAEL BEHE, DARWIN'S BLACK BOX (1998).
182. MILLER, *supra* note 180, at 129–62.

7. The Madison Avenue Approach to Law and Science

1. Ronald L. Numbers, The Creationists: From Scientific Creationism to Intelligent Design 7–8 (exp. ed. 2006).
2. *Id.*
3. *See generally id.*
4. *Id.* at 8, 92–119, 368–72.
5. *See generally id.*
6. William Paley, Natural Theology (paperback ed. 1963).
7. Numbers, *supra* note 1, at 7–8.
8. Edward J. Larson, Summer for the Gods (1997).
9. Epperson v. Arkansas, 393 U.S. 97, 106–09 (1968).
10. *See generally* Numbers, *supra* note 1.
11. *Id.*
12. *Id.*
13. *Id.* at 292–94.
14. *Id.* at 7–8.
15. *Id.* at 55–64.
16. *Id.* at 259, 266–68, 291–92, 307–08.
17. *Id.* at 88–106, 135, 349.
18. *Id.* at 116–18, 260.
19. *Id.* at 116.
20. *Id.*
21. *Id.*
22. Robert Pennock, Tower of Babel: The Evidence against the New Creationism (1999).
23. Numbers, *supra* note 1.
24. Discovery Institute, Wedge Document (1998), http://www.antievolution.org/features/wedge.html; *see also* Barbara Forrest & Paul R. Gross, Creationism's Trojan Horse (2004) (discussing the Wedge Document in detail).
25. Plato, Timaeus (Peter Kalkavage trans., 2001); Cicero, De Natura Deorum I (Richard McKirahan ed., 1997); Thomas Aquinas, The Summa Theologica of St. Thomas Aquinas (Christian Classics 1981); Paley, *supra* note 6.
26. Isaiah Berlin, The Age of Enlightenment (1984).

27. THE CHURCH AND GALILEO (Ernan McMullin ed., 2005).
28. INGRID D. ROWLAND, GIORDANO BRUNO: PHILOSOPHER/HERETIC (2009).
29. ROWLAND, *supra* note 28; CHURCH AND GALILEO, *supra* note 27.
30. *See supra* note 15, and accompanying text.
31. LARSON, *supra* note 8; NUMBERS, *supra* note 1.
32. CHARLES DARWIN, THE AUTOBIOGRAPHY OF CHARLES DARWIN (Nora Barlow ed., 1993).
33. LARSON, *supra* note 8; PENNOCK, *supra* note 22.
34. NUMBERS, *supra* note 1.
35. NUMBERS, *supra* note 1, at 7–8; PENNOCK, *supra* note 22.
36. PENNOCK, *supra* note 22.
37. *See* Chapters 1 and 3.
38. Frank S. Ravitch, *Playing the Proof Game: Intelligent Design and the Law*, 113 PENN. ST. L. REV. 841 (2009).
39. PHILLIP JOHNSON, DARWIN ON TRIAL (1991).
40. *Id.*
41. Stephen Jay Gould, *Impeaching a Self Appointed Judge*, 267 SCIENTIFIC AMERICAN 118–121 (July 1992).
42. *See, e.g.*, Raymond Bohlin, *Review of Darwin on Trial*, PROBE MINISTRIES (1992) (short review indicative of other reviews in evangelical publications praising the book), http://www.leaderu.com/orgs/probe/docs/darwin.html.
43. FORREST & GROSS, *supra* note 24, at 15–23.
44. *Id.*
45. *Id.*
46. *Id.*
47. *Id.*
48. *Id.*
49. *See generally id.*; Ravitch, *supra* note 38; Kitzmiller v. Dover Area School District, 400 F.Supp.2d 707, 735–46 (E.D. Pa. 2005).
50. Wedge Document, *supra* note 24; FORREST & GROSS, *supra* note 24.
51. *See* Chapters 1 and 3.
52. *Id.*
53. PENNOCK, *supra* note 22.

54. *Id.*
55. Robert T. Pennock, *Can't Philosophers Tell the Difference between Science and Religion?: Demarcation Revisited*, SYNTHESE (April 2009).
56. *See* Chapters 1 and 3.
57. *Id.*
58. *Id.*; *see also* Wedge Document, *supra* note 24; FORREST & GROSS, *supra* note 24.
59. *See* Chapters 1 and 4; *see also* note 25, and accompanying text.
60. Edwards v. Aguillard, 482 U.S. 578 (1987).
61. Ravitch, *supra* note 38; FRANK S. RAVITCH, MASTERS OF ILLUSION: THE SUPREME COURT AND THE RELIGION CLAUSES (2007).
62. Kitzmiller, 400 F.Supp.2d at 745.
63. Pennock, *supra* note 55.
64. *Id.*
65. Kitzmiller, 400 F.Supp.2d 707; Jane Gitschier, *Taken to School: An Interview with the Honorable Judge John E. Jones, III*, PLOS GENETICS (Dec. 5, 2008).
66. Anika Smith, *What NOVA Won't Tell You about Dover*, Discovery Institute, Center for Science and Culture (Nov. 13, 2007), http://www.discovery.org/a/4300; David K. DeWolf, John G. West, & Casey Luskin, *Intelligent Design Will Survive Kitzmiller v. Dover*, 68 MONT. L. REV. 7 (2007); Michael Behe, *Whether Intelligent Design Is Science*, Discovery Institute, Center for Science and Culture (Feb. 3, 2006), http://www.discovery.org/a/3218.
67. *See id.*
68. *Id.*
69. *Id.*
70. *Id.*
71. Discovery Institute, Center for Science and Culture, *Setting the Record Straight about Discovery Institute's Role in the Dover School District Case* (Nov. 10, 2005), http://www.discovery.org/a/3003.
72. Pennock, *supra* note 55, at *Introduction*.
73. *Id.*
74. *Id.*

75. *Id.*
76. *See infra* note 80.
77. *Id.*
78. *Id.*
79. Kitzmiller, 400 F.Supp.2d 707.
80. Pennock, *supra* note 55.
81. Kitzmiller, 400 F.Supp.2d at 721–22.
82. This was exemplified by the response of certain religious communities to Ben Stein's movie EXPELLED: NO INTELLIGENCE ALLOWED (Premise Media Corp. 2008).
83. *See* DeWolf et al., *supra* note 66.
84. *See id.*; *see also* Jay D. Wexler, *Kitzmiller and the "Is It Science?" Question*, 5 FIRST AMEND. L. REV. 90 (2006).
85. *See, e.g.*, *id.*
86. *See id.*; *see also* Kelly S. Terry, *Shifting Out of Neutral: Intelligent Design and the Road to Nonpreferentialism*, 18 B.U. PUB. INT. L.J. 67 (2008); Nicholas A. Schuneman, *One Nation, Under . . . the Watchmaker?: Intelligent Design and the Establishment Clause*, 22 BYU J. PUB. L. 179 (2007).
87. *See, e.g.*, Caprice L. Roberts, *In Search of Judicial Activism: Dangers of Quantifying the Qualitative*, 74 TENN. L. REV. 567 (2007); Arthur D. Hellman, *Judicial Activism: The Good, the Bad, and the Ugly*, 21 MISS. L. REV. 253 (2002).
88. *See, e.g.*, *id.*
89. *See* Chris Dixon, *Think Tank Fuels Debate on Evolution*, POST AND COURIER (Charleston, S.C.) (March 5, 2006), at A1; Eyana Adah McMillan, *Rehm New Board Member in Dover*, YORK DISPATCH (York, P.A.) (Jan. 4, 2006); Michelle Starr, *Intelligent Design Supporters Ousted*, EVENING SUN (Hanover, P.A.) (Nov. 9, 2005), at A1.
90. *See* Dennis Overby, *Philosophers Notwithstanding, Kansas School Board Redefines Science*, N.Y. TIMES (Nov. 15, 2005), at F3; Peter Slevin, *Kansas Board of Education First to Back "Intelligent Design"; Schools to Teach Doubts about Evolutionary Theory*, WASHINGTON POST (Nov. 9, 2005), at A1.

91. *See* Slevin, *supra* note 90; *see also* Chris Moon, *Science Revisited*, TOPEKA-CAPITAL JOURNAL (Nov. 9, 2005), at A1.
92. *See id.*; *see also* Jodi Wilgoren, *Kansas School Board Moves to Challenge Evolution – Again/Religious Backers of "Intelligent Design" Seek to Change How Science Is Taught*, N.Y. TIMES (May 6, 2005).
93. *See Religion and the Public Schools; Charles Darwin Prevails in Kansas, as in Dover, but the Debate Will Persist*, MORNING CALL (Allentown, P.A.) (Aug. 8, 2006), at A8; Slevin, *supra* note 90; *Stealth Attack on Evolution; Who Is behind the Movement to Give Equal Time to Darwin's Critics, and What Do They Really Want?*, TIME (Jan. 31, 2005), at 53.
94. *See id.*; *see also* Kitzmiller, 400 F.Supp.2d at 721.
95. *See* Linda Shaw, *Seattle Think Tank Raises Questions about Evolution*, SEATTLE TIMES (April 5, 2005); John Hann, *"The Big E-word": The Kansas State Board of Education Is Poised to Begin Another Round of Debates over Teaching Creationism in Public School Science Classes*, SAINT PAUL PIONEER PRESS (April 27, 2005), at A3; *Some Groups to Boycott Kansas Hearings on Evolution Education*, EDUCATION WEEK (April 27, 2005), at 3.
96. *See* Chapters 1, 3, 4, and 6.
97. *See supra* note 99; *see also* Lisa Anderson, *Darwin's Theory Evolves into a Culture War; Kansas Curriculum Is Focal Point of Wider Struggle across Nation*, CHICAGO TRIBUNE (May 22, 2005), at C1; P. J. Huffstutter, *Evolution Isn't a Natural Selection Here; Kansas Looks Again at Whether Teachers Should Be Allowed to Present Non-scientific Theories*, LOS ANGELES TIMES (May 6, 2005), at A1; David Klepper, *Kansas School Board's Hearings on Evolution End in Acrimony*, KANSAS CITY STAR (May 13, 2005).
98. *See id.*; *see also Some Groups to Boycott Kansas Hearings on Evolution Education*, EDUCATION WEEK (April 27, 2005), at 3.
99. *See, e.g.*, *id.*; *see also* Josh Funk, *Kansas Board Members Endorse Criticisms of Evolution*, WICHITA EAGLE (June 10, 2005); Jodi Wilgoren, *In Kansas, Darwinism Goes on Trial Once More*, N.Y. TIMES (May 6, 2005), at A18; *Kansas Hears from Critics of*

Evolution; Foes Dominate Sessions before State Board Panel, EDUCATION WEEK (May 11, 2005), at 1.

100. Discovery Institute, Center for Science and Culture, *Discovery Institute Statement on the Kansas Science Standards Situation* (Aug. 1, 2006), http://www.discovery.org/a/3683.
101. *See, e.g.*, *Evolution Loses and Wins, All in One Day; Kansas Adopts Standards Critical of Topic; Dover, Pa., Voters Dump "Intelligent Design" Board*, EDUCATION WEEK (Nov. 16, 2005), at 1; David Klepper, *Kansas Board Approves New Science Standards*, KANSAS CITY STAR (Nov. 9, 2005).
102. *See* Charles Krauthammer, Editorial, *Phony Theory, False Conflict; "Intelligent Design" Foolishly Pits Evolution against Faith*, WASHINGTON POST (Nov. 18, 2005), at A23; Mike Lafferty, *Intelligent Design; Like Ohio, Kansas Redefined "Science,"* COLUMBUS DISPATCH (Nov. 17, 2005), at D5.
103. *See* Monica Davey & Ralph Blumenthal, *Fight over Evolution Shifts in Kansas School Board Vote*, N.Y. TIMES (Aug. 3, 2006), at A15; *Conservatives Lose in Kansas Board Vote*, THE TIMES UNION (Albany, N.Y.) (Aug. 3, 2006), at A5; Nicholas Riccardi, *Evolution Foes Lose Their Edge on Kansas Board; The State Education Panel Appears to Be Split 6–4 between Supporters of Darwin's Theory and Advocates of Teaching Intelligent Design*, LOS ANGELES TIMES (Aug. 2, 2006), at A19.
104. *See Not in Kansas Anymore; Just When Kansas Returns Science to the Classroom, Rep. Chisum Tries to Goad Texas toward Dark Ages*, HOUSTON CHRONICLE (Feb. 16, 2007), at B10; *Darwin Back on Kansas' A List; The State Board of Education Votes against Guidelines Hostile to Teaching Evolution*, LOS ANGELES TIMES (Feb. 14, 2007), at A19.
105. EXPELLED: NO INTELLIGENCE ALLOWED (Premise Media Corp. 2008).
106. *See, e.g.*, *Hannity and Colmes: Movie Challenges Scientific Theory* (Fox News Network television broadcast, April 11, 2008); *The Big Story with John Gibson: Interview with Discovery Institute's William Dempsky* (Fox News Network television broadcast, Aug.

2, 2005); *The O'Reilly Factor: The Controversy over Intelligent Design* (Fox News Network television broadcast, Dec. 23, 2004).

107. 2007 I.R.S. Form 990, Return of Organization Exempt from Income Tax, Discovery Institute (filed November 21, 2008) (showing total direct public support for 2007 as $3,998,390); 2006 I.R.S. Form 990, Return of Organization Exempt from Income Tax, Discovery Institute (filed November 21, 2007) (showing total direct public support for 2006 as $3,781,988).
108. Edwards v. Aguillard, 482 U.S. 578 (1987).
109. Discovery Institute, Center for the Renewal of Science and Culture, *The Wedge*, http://ncse.com/webfm_send/747.
110. *See* http://biologicinstitute.org/about/.
111. NUMBERS, *supra* note 1.
112. *Id.* at 310–11.
113. *Id.*
114. *Id.*
115. PENNOCK, *supra* note 22.
116. *Id.*
117. *Id.*
118. Kitzmiller, 400 F.Supp.2d 707.
119. *See* http://biologicinstitute.org/about/.
120. *Id.*
121. *Id.*
122. Biologic Institute Web site, http://www.biologicinstitute.org.
123. For a more detailed discussion of William Paley, the watchmaker God, and natural theology, *see* Chapter 2 and this chapter, Section I.
124. Biologic Institute, *Our Take on the ID Controversy*, http://biologicinstitute.org/our-take-on-the-id-controversy.
125. EDWARD O. WILSON, FROM SO SIMPLE A BEGINNING: DARWIN'S FOUR GREAT BOOKS (2005).
126. *Id.*
127. *Id.*
128. *Id.*, *repr. in Can Biology Do Better Than Faith?*, NEW SCIENTIST (Nov. 2005).

129. *Id.*
130. *Id.*
131. *Id.*
132. Kitzmiller, 400 F.Supp.2d at 736.
133. GRAHAM BELL, SELECTION: THE MECHANISM OF EVOLUTION (1997).
134. *Id.*
135. *Id.* at xvii–xviii (emphasis added).
136. *Id.* at xviii.
137. *Id.*
138. *See* Chapter 5.
139. *See* Chapters 1–4.
140. WILSON, *supra* note 128; *see also* Chapter 1.
141. *See* Chapters 1 and 5.
142. *Id.*
143. WILSON, *supra* note 128; RICHARD DAWKINS, THE GOD DELUSION (2006).
144. PENNOCK, *supra* note 22; Kitzmiller, 400 F.Supp.2d 707.
145. *Id.*; *see also* Chapters 1 and 2.
146. *Id.*
147. *See generally* KENNETH R. MILLER, FINDING DARWIN'S GOD (2002).
148. AL RIES & JACK TROUT, MARKETING WARFARE (1997).
149. *Id.*
150. *Id.*
151. *Id.*
152. *Id.*
153. *Id.*
154. Edwards v. Aguillard, 482 U.S. 578 (1987).
155. 505 U.S. 577 (1992).
156. Stephen B. Pershing, *Graduation Prayer after Lee v. Weisman: A Cautionary Tale*, 46 Mercer L. Rev. 1097, 1100 (1995).
157. *Id.* at 1100–01.
158. *Id.*
159. *Id.*

160. Lee, 505 U.S. at 595–96; Everson v. Bd. of Education of Ewing Tp., 330 U.S. 1 (1947).
161. 374 U.S. 203 (1963).
162. 370 U.S. 421 (1962).
163. Lee, 505 U.S. 577.
164. *Id.*
165. 930 F.2d 416 (5th Cir. 1991) (*Jones I*), *vacated and remanded*, 112 S. Ct. 3020 (1992), *on remand*, 977 F.2d 963 (5th Cir. 1992) (*Jones II*), *cert. denied*, 113 S.Ct. 2950 (1993). Unless otherwise specified in the text, all references to the *Jones* case are to *Jones II*.
166. 930 F.2d 416 (5th Cir. 1991) (*Jones I*).
167. *Id.*
168. *See* Jones II, 977 F.2d at 964.
169. Jones v. Clear Creek Indep. Sch. Dist., 112 S.Ct. 3020 (1992); Jones II, 977 F.2d at 965.
170. Jones II, 977 F.2d 963.
171. *See, e.g.*, Black Horse Pike, 84 F.3d 1471; Harris, 41 F.3d 447 (9th Cir. 1994), *vacated as moot*, 115 S.Ct. 2604 (1994); Jonathan C. Drimmer, *Hear No Evil, Speak No Evil: The Duty of Public Schools to Limit Student-Proposed Graduation Prayers*, 74 Neb. L. Rev. 411 (1995); E. Gregory Wallace, *When Government Speaks Religiously*, 21 Fla. St. U. L. Rev. 1186, 1261–62 (1994).
172. Jones II, 977 F.2d at 966–68.
173. *Id.* at 968–69.
174. Jones II, 977 F.2d 963.
175. *Id.* at 970.
176. *Id.* at 970–71.
177. *Id.* at 971.
178. *Id.* at 971–72.
179. *Id.* at 971.
180. Frank S. Ravitch, School Prayer and Discrimination 57 (2001).
181. *Id.*
182. *Id.*
183. Ingebretsen v. Jackson Public School District, 88 F.3d 274, 278 (5th Cir.1996).

184. Santa Fe Indep. Sch. Dist. v. Doe, 530 U.S. 290 (2000).
185. Pershing, *supra* note 157, at 1100.
186. RAVITCH, *supra* note 181, at 57.
187. Edwards v. Aguillard, 482 U.S. 578 (1987); Epperson v. Arkansas, 393 U.S. 97, 106–09 (1968).
188. James McKinley Jr., *In Texas, a Line in the Curriculum Revives Evolution Debate*, NEW YORK TIMES (Jan. 21, 2009).
189. See Chapters 3 and 6.
190. Kim Kozlowski, *Evolution Battle Grows in Schools*, DETROIT NEWS (July 24, 2005) (discussing the situation in Gull Lake, Michigan, where teachers were prevented from teaching ID in science classes and a lawsuit was threatened).
191. *Id.*

Conclusion

1. Edwards v. Aguillard, 482 U.S. 578 (1987).
2. *See* Chapter 3.

Index

Index

Index

Index